AF714104

Evaluating and Testing the COKING PROPERTIES OF COAL

RAYMOND E. ZIMMERMAN

A WORLD COAL BOOK

MILLER FREEMAN PUBLICATIONS, INC.
San Francisco

Copyright © 1979 by the Paul Weir Company

Library of Congress Catalog Card Number: 79-84399
International Standard Book Number: 0-87930-111-2

All rights reserved. No part of this book covered by the copyrights hereon may be reproduced or copied in any manner whatsoever without written permission except in the case of brief quotations embodied in articles and reviews. For information contact the publishers, Miller Freeman Publications, Inc., 500 Howard Street, San Francisco, California 94105, USA.

Printed in the United States of America.

Contents

List of figures

List of tables

Foreword

In connection with the Paul Weir Company's work as worldwide consultants on coal, it became apparent that there is a need for a concise manual describing various methods for evaluating the coking properties of coals and discussing the important tests used in making this evaluation.

It has been our experience that there is no one source of information on this subject; rather, data on coking coals come from many sources, i.e., various societies' proceedings, technical magazines, and textbooks on certain specialized areas. The amount of information available would fill a library.

In 1961, the Paul Weir Company prepared a report of a similar title. The demand for it exceeded all expectations, and numerous copies had to be reproduced at various times in the ensuing period. Much of the information incorporated in the original text still stands; however, obviously, in the years since its original inception, there have been many changes in processes for determining the coking properties of coals, and new test procedures have been developed.

As a result, this manual has been prepared. We are the first to admit that it is not all-inclusive. The knowledge of coking coal characteristics is expanding. The reader may note omissions of material he feels should have been included; we welcome the reader's suggestions for additional information.

The text and information presented in this edition have been authored by Raymond E. Zimmerman, a vice president of the Paul Weir Company. Mr. Zimmerman is a worldwide authority on the subject, as well as in the field of coal preparation.

We hope that this manual will be helpful to the many coal producers or coal industry interests who do not have in-depth expertise concerning coking coals but who wish to obtain some background knowledge of what is required in the coking coal market. It is hoped that this information will aid their evaluation.

John P. Weir
President
Paul Weir Company
Chicago, Illinois
January 1979

Preface

The main purpose of this manual is to provide the reader with a ready reference for the various standards involved in evaluating coking coals, including the coal classification standards used by the major exporting and importing countries.

It is not our intention to describe in exact detail the numerous official test procedures adopted by the standards organizations of the various countries, including the U.S.A. This detailed information can and should be obtained from the governments and national societies involved. An extensive, but by no means all-inclusive, bibliography is included, which we believe contains some of the most important references.

Since certain standards can be difficult to procure, the appendices to this manual include those which many find difficult to obtain.

It is to be remembered that standards change from time to time, or they may be modified to varying degrees. Some standards may be tentative. It is well for those involved in such tests to make sure they are using the latest procedure.

The principal American sources for standards information are the various publications of the U.S. Bureau of Mines [1] and the latest annual standards of the American Society for Testing and Materials, Committee D-5 [2]. For Europe, certain tests recommended and described by the International Organization for Standardization (ISO), an agency of the United Nations Economic Commission for Europe [3] are in common use.

Individual countries frequently have their own coking coal standards, and coal producers interested in selling coal to these countries should obtain their particular standards [4, 5].

We do not claim that this manual is a definitive textbook covering all the technical details and scientific data required by a specialist in coking coal technology. Such a project would require many volumes. Rather, we hope it will serve as a guide to acquaint nonspecialists with the things they should know to determine the coking properties of their coals and to give them some idea as to the value of their coals in the coking coal market.

Raymond E. Zimmerman
Vice President
Paul Weir Company
Chicago, Illinois
January 1979

1 The characteristics of coking coals

Coal is a heterogeneous, carbonaceous, fossilized material of almost infinite variations in consist—from low-grade lignite or brown coal to graphitic-like meta-anthracite. Among its many properties is the ability of some types of coal to form coke when heated in ovens or retorts. During the application of varying degrees of temperature in an atmosphere which will not support combustion, particles of some types of coals will first coalesce or melt, giving off volatile gases and vapors; will then become fluid; and finally will harden into a porous or cellular substance called *coke*. Since it is high in fixed carbon, coke is valuable as a fuel and as a reducing agent in many metallurgical and chemical processes.

There are wide differences of opinion as to which characteristics or combinations of properties determine that some coals will be of suitable coking quality while others will not be. A large amount of literature has been published concerning the causes of this phenomenon, and there is still much to be learned. Most of the tests to determine coking properties are essentially empirical, and no specific test tells the complete story.

Coking coals, by definition, are coals that when heated to sufficiently high temperatures pass through a transient plastic stage in which they successively soften, swell, and resolidify into a coherent cellular coke. The fact that coking coals have this property enables investigators to apply methods of measuring the plastic range of coals to determine their coking potential. Figure 1 portrays an idealized curve of the plastic and resolidification range. The interval (in terms of degrees) between onset and termination of plasticity is normally on the order of onset at from 50° to 75°C, and termination (somewhat dependent on rank) somewhere between 350° and 500°C.

The ideal pattern of softening, swelling, and resolidification is not always evident in actual practice. In many coals, particularly those of very high volatile content, swelling may be almost completely masked by bulk shrinkage as a result of loss of volatile matter.

In addition to varying in general with the rank of the coal, all phases of the plastic state necessary for the proper development of coke may vary for the *same* coal by the rate at which the sample is heated, by prolonged oxidation (or weathering) of the sample, by mean particle size, and by bulk density. The plastic properties of some coals can even be destroyed by heat treatment at temperatures just below the softening point. In fact, one method of controlling

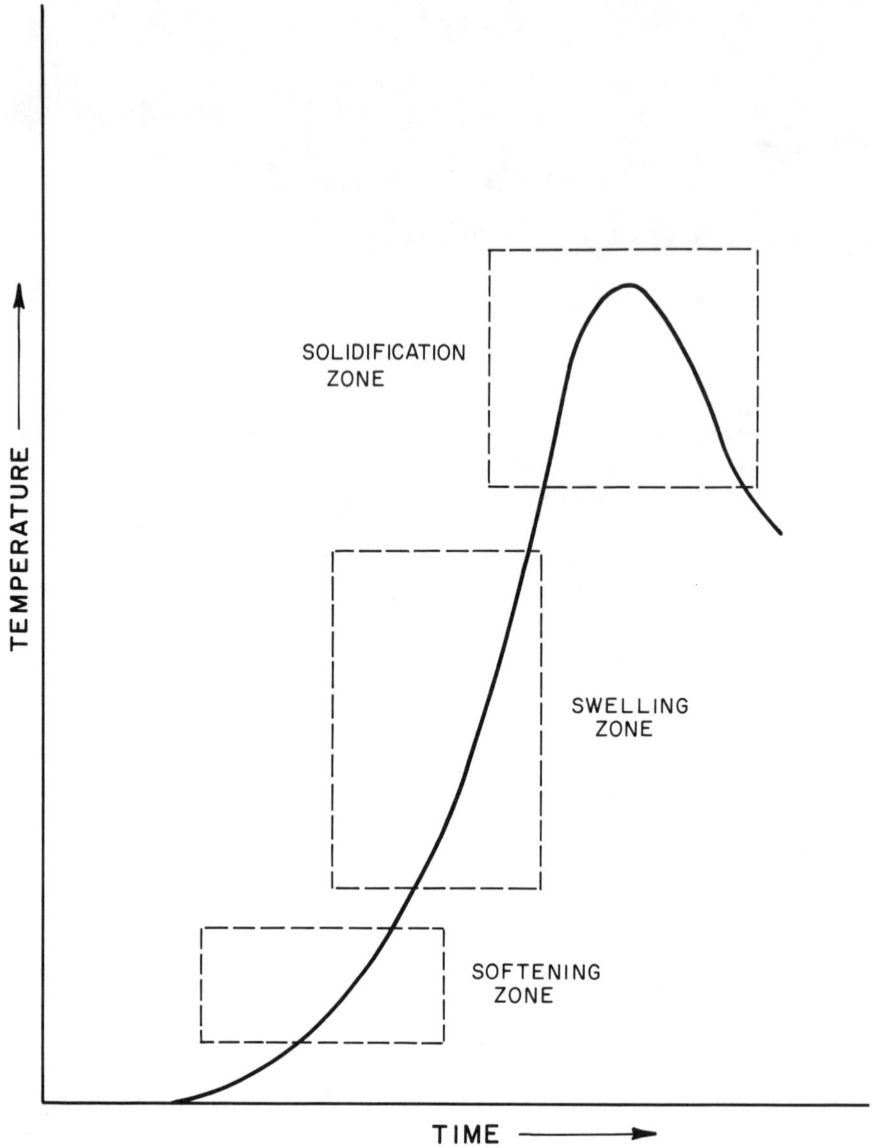

Figure 1. Plastic zones of coking coals.

a highly plastic coal (if it is desirable to change its plasticity) is by preoxidation or heat treatment before coking. Some coals, on the other hand, may appear inferior under slow rates of heating but may produce a satisfactory coke under more rapid heating rates.

In evaluating the coking properties of a coal, it is important that the condition or treatment of the sample be completely known. A fresh, unexposed sample of coal may possess suitable coking properties, but if tested after oxidation through exposure to the atmosphere for various lengths of time, may contrarily exhibit negative or inferior results. Accordingly, samples of freshly mined coal should be shipped to laboratories in airtight containers. Coal in place near its outcrop, or under shallow or unconsolidated overburden, may also be too weathered to reveal its true coking characteristics. Such weathering processes, however, seldom penetrate a firm coal bed more than 30 to 50 feet from the outcrop or surface and are usually denoted or accompanied by earthy stains along joint or cleavage surfaces within the coal bed. In any case, this must be checked to ensure a representative sample.

Another factor that must be taken into consideration in testing a sample of coal for its coking properties is the amount of extraneous impurities (such as shale partings or other non-coal material) it may contain. The presence of notable amounts of such extraneous materials may be sufficient to lower or destroy the natural coking characteristics of the coal. Thus, a necessary adjunct to preparing a sample for carbonization tests may be the removal of these impurities, at least to the extent that they can be removed by normal coal-washing procedures.

Although coal rank, as expressed by fixed carbon contents or related parameters, can be used as a general guideline for identifying the range of coking coals, it is not always a determining factor. While it is true that North American and western European coking coals are almost entirely confined to those having carbon limits between 69 and 78 percent on a dry, ash-free basis (and are accordingly classifiable as low-, medium-, or high-volatile bituminous coals), a wider study of coal systematics suggests that this relationship may be relatively coincidental.

One common factor, aside from plastic properties, that all coking coals possess regardless of rank is that they tend to be markedly compact and that their moisture contents, porosities, and internal surface areas are low in comparison with non-coking coals. Just where the limit of bed (or inherent) moisture content may be, however, is difficult to say. Illinois coking coals may have as high as 10 percent bed moisture. The coking subbituminous coals near Tunçbilek, Turkey, have a moisture content of from 14 to 16 percent.

One thing that stands out in evaluating coking coals is that there is no single index number that can be used to place a positive value on the coking properties of a particular coal. Chemical analyses are just as important as the various physical tests described.

The one combination of physical tests that may be used with a greater degree of confidence than the others is the Composition Balance Index (CBI) and the Strength Index (SI), determined by petrographic analyses. When plotted, these can give a coke strength or stability value; however, the test is generally done on a blend of coals. But even this combination test can be subject to various errors of measurement in petrographic analyses.

Other important physical tests, particularly in single coals, are Gieseler fluidity; free-swelling index; various dilatometer tests; Gray-King; Roga; and tumbler tests, such as the ASTM, Micum, or JIS. All are described in this manual.

Chemical analyses—including ash, heating value, volatile matter, sulfur, phosphorus, and the fusion temperature and mineral analyses of the ash—are equally important. The purpose of using coke is to provide carbon for reduction of ore, and this coke must be of the proper strength and structure to give maximum production of metal, as well as contain a minimum of deleterious non-carbon minerals.

It should be emphasized that very few coals are used singly to make coke; practically all steel producers use blends, often of widely differing coals. So, the blending value of a coal is highly important. This could mean that a particular coal could have more value to some users than to others, depending on the other types of coals they must use.

The effect of blending coals of different characteristics is shown in many tests described in this manual. For specific information on coal blends, refer to Table 4 and Figures 9, 12, 15, 17, and 24, which show the effect of various coal blends on pertinent coking coal characteristics.

Few users of coking coals, when dealing with an unknown coal, will rely solely on the chemical and physical tests described in this text, but will insist on at least some small-scale coke oven testing; they will often require a large-scale shipment to test in commercial-sized ovens before final purchase.

Finally, it should again be emphasized that the accuracy of the coal sample itself is of vital importance. In other words, does the sample truly represent the coal to be produced?

Although not part of the subject matter outlined in this manual, it should certainly be mentioned that coke producers may employ many additional procedures to enhance the quality of the coke and the capacity of their coke ovens. Factors and variables include: increase of bulk density in the ovens (essential for capacity, heating rates, and improving coking characteristics); control of the degree of pulverization of coal before charging; use of additives, such as oil, pitch, coke breeze, and briquets; variable heating rates; coking time; and possibly preheating of the coal before charging into the ovens.

There is increasing interest in the addition of coal briquets made from coal having inferior coking properties to coke oven charges, to minimize the use of high-quality (and higher priced) coking coals. The Japanese steel industry is advanced in this practice, with claims of using as much as 30 percent of the charge in the form of coal briquets.

That practice is not to be confused with the use of "formed coke," an approach in which relatively non-coking coal is briquetted, carbonized, and used directly in blast furnaces.

Whether these methods are used depends entirely upon the technology of the processor and the availability of high-quality coking coal at an acceptable price. It is chiefly a matter of economics.

Although authorities differ in estimates of coking coal requirements in the U.S.A., there is general agreement that about 100 million tons of coking (metallurgical) coal will be consumed by 1980 [44]. This estimate could be altered somewhat by a breakthrough in "formed coking" techniques.

It is estimated that world demand for coking coal may reach 681.5 million tons by 1985. In 1970, U.S. coking coal exports were approximately 12.7 percent of foreign requirements. By 1985, the U.S. export market could reach 20 percent of world requirements. This will depend, however, upon many factors, economic as well as technological.

The greatest potential increase in coking coal demand appears to be in newly developing countries, where, as industrialization advances, an increasing need for steel, and therefore coke, is generated. Furthermore, these areas of the world are generally deficient in indigenous coking coal sources.

The largest importer of coking coal of any of the advanced industrialized nations is Japan, which has relatively small sources of its own. Despite the current slowing down of its steel industry, its need for high-grade coking coal in large tonnages will be a fact for the foreseeable future.

It is our opinion that the need for coking coal will always be substantial, and that the knowledge of coking coal characteristics is vital to those wishing to enter the market or continue their coking coal markets.

2 The classification of coals

The extremely wide variety of coals known in this world makes it essential that they be classified according to systems or rank. Although there are many marginal cases where such classification is difficult (and, as previously mentioned with respect to coking properties, should be used with caution), coals generally can be identified in accordance with some systematic scale. Therefore, in analyzing coals, an attempt should be made to place them for comparative purposes in the current national or international classifications.

Although many of the major coal-producing countries have their own classification schemes, the engineer or geologist needs generally to be concerned with but two major systems. One is the American system [1, 2]; the other is the international classification system proposed by the International Organization for Standardization (ISO) [3], an agency of the United Nations Economic Commission for Europe. Because of their importance, both systems are reviewed in this manual. The British Standards [4] are given in Table 3, at the end of this chapter. The Australian system is similar to the ISO, but also includes ash levels. The Japanese generally refer to ASTM, but also have their Japanese Industrial Standards (JIS) [5].

AMERICAN (ASTM) CLASSIFICATION OF COALS BY RANK

The American system [1, 2], designated ASTM D-388-66, was adopted by the American Society for Testing and Materials (1958), and has also been adopted by the American Standards Association (ASA) as ASA No. M 20.1-1973. (Table 1 shows the ASTM classifications.) The basic scheme of classification is related to fixed carbon and Btu calculated to the mineral-matter-free basis. The higher ranked coals are classified according to fixed carbon on a dry basis; and the lower rank coals, according to Btu on a moist basis. Agglomerating and weathering properties are used to differentiate between certain adjacent groups.

This classification does not include a few coals, principally non-banded varieties, that have unusual physical and chemical properties and that fall within the limits of fixed carbon or calorific value of the high-volatile bituminous and subbituminous ranks. All these coals either contain less than 48 percent carbon on a dry, mineral-matter-free basis or have more than 15,500 British thermal units per pound (Btu/lb) on a moist, mineral-matter-free basis.

Moist basis refers to natural inherent moisture, and does not include extraneous surface water. It is sometimes referred to as *air dry* basis, a term frequently used in standard laboratory analyses.

Table 1. ASTM Classification of Coals by Rank

Class	*Group*	*Analytical Limits (Mineral-Matter-Free Basis)*	*Requisite Physical Properties*
I. Anthracite	1. Meta-Anthracite	Dry FC 98% or more; Dry VM 2% or less.	
	2. Anthracite	Dry FC 92% or more and less than 98%; Dry VM 8% or less and more than 2%.	
	3. Semianthracite	Dry FC 86% or more and less than 92%; Dry VM 14% or less and more than 8%.	Non-agglomerating
II. Bituminous	1. Low-Volatile	Dry FC 78% or more and less than 86%; Dry VM 22% or less and more than 14%.	Agglomerating commonly
	2. Medium-Volatile	Dry FC 69% or more and less than 78%; Dry VM 31% or less and more than 22%.	
	3. High-Volatile A	Dry FC less than 69%; Dry VM more than 31% and moist Btu 14,000 or more.	
	4. High-Volatile B	Moist Btu 13,000 or more and less than 14,000.	
	5. High-Volatile C	Moist Btu 10,500 or more and less than 13,000.	Either agglomerating or non-agglomerating
III. Subbituminous	1. Subbituminous A	Moist Btu 10,500 or more and less than 11,500.	Non-agglomerating
	2. Subbituminous B	Moist Btu 9,500 or more and less than 10,500.	
	3. Subbituminous C	Moist Btu 8,300 or more and less than 9,500.	
IV. Lignitic	1. Lignite A 2. Lignite B	Moist Btu less than 8,300. Moist Btu less than 6,300.	Non-agglomerating

Source: ASTM D-388-66 (reprinted by permission).

Legend: FC = Fixed Carbon
VM = Volatile Matter
Btu = British Thermal Units

Note: 429.923 Btu = 1 MJ/kg (megajoules), specific energy, SI.

GROUPS (determined by caking properties)			CODE NUMBERS										SUBGROUPS (determined by coking properties)		
GROUP NUMBER	ALTERNATIVE GROUP PARAMETERS: Free-swelling index (crucible-swelling number)	ALTERNATIVE GROUP PARAMETERS: Roga Index	The first figure of the code number indicates the class of the coal, determined by volatile matter content up to 33% V.M. and by calorific parameter above 33% V.M. The second figure indicates the group of coal, determined by caking properties. The third figure indicates the subgroup, determined by coking properties.										SUB-GROUP NUMBER	ALTERNATIVE SUBGROUP PARAMETERS: Dilatometer	ALTERNATIVE SUBGROUP PARAMETERS: Gray-King
3	>4	>45					435	535	635				5	>140	>G8
						334	434	534	634				4	>50-140	G5-G8
						333	433	533	633	733			3	>0-50	G1-G4
						332 a, 332 b	432	532	632	732	832		2	≤0	E-G
2	2½-4	>20-45				323	423	523	623	723	823		3	>0-50	G1-G4
						322	422	522	622	722	822		2	≤0	E-G
						321	421	521	621	721	821		1	Contraction only	B-D
1	1-2	>5-20			212	312	412	512	612	712	812		2	≤0	E-G
					211	311	411	511	611	711	811		1	Contraction only	B-D
0	0-½	0-5		100 (A, B)	200	300	400	500	600	700	800	900	0	Nonsoftening	A
CLASS NUMBER →			0	1	2	3	4	5	6	7	8	9	As an indication the following classes have an approximate volatile-matter content of: Class 6 33-41% volatile matter; 7 33-44% " "; 8 35-50% " "; 9 42-50% " "		
CLASS PARAMETERS	Volatile matter (dry, ash-free) →		0-3	>3-10 (A: >3-6.5; B: >6.5-10)	>10-14	>14-20	>20-28	>28-33	>33	>33	>33	>33			
	Calorific parameter a/ →		–	–	–	–	–	–	>13,950	>12,960-13,950	>10,980-12,960	>10,260-10,980			
CLASSES (Determined by volatile matter up to 33% V.M. and by calorific parameter above 33% V.M.).															

Note: (i) Where the ash content of coal is too high to allow classification according to the present system, it must be reduced by laboratory float-and-sink method (or any other appropriate means). The specific gravity selected for flotation should allow a maximum yield of coal with 5 to 10 percent ash.
(ii) 332a...>14-16% V.M.
332b...>16-20% V.M.

a/ Gross calorific value on moist, ash-free basis (30°C., 96% relative humidity) B. t. u./ lb.

Figure 2. International classification of hard coals by type. From Reference 7.

Determinations of mineral-matter-free fixed carbon, volatile mater, and Btu/lb for placement in the ASTM classification are made in accordance with the following formulas:

$$\text{Dry, Mm-free FC} = \frac{\text{FC} - 0.15\ \text{S}}{100 - (\text{M} + 1.08\ \text{A} + 0.55\ \text{S})} \times 100$$

$$\text{Dry, Mm-free VM} = 100 - \text{Dry, Mm-free FC}$$

$$\text{Moist, Mm-free Btu} = \frac{\text{Btu} - 50\ \text{S}}{100 - (1.08\ \text{A} + 0.55\ \text{S})} \times 100$$

where:

Mm = Mineral matter
Btu = British thermal units (specific energy)
FC = % fixed carbon
VM = % volatile matter
M = % moisture
A = % ash
S = % sulfur

Note: "Moist" Btu refers to coal containing its natural-bed, or inherent, moisture but not surface moisture.

Symbols often used in designating rank, in accordance with the ASTM classification, are:

Lvb = Low-volatile bituminous
Mvb = Medium-volatile bituminous
Hvab = High-volatile A bituminous
Hvbb = High-volatile B bituminous
Hvcb = High-volatile C bituminous
Sub a = Subbituminous A
Sub b = Subbituminous B
Sub c = Subbituminous C

Detailed explanations for determining coal rank are, of course, contained in the ASTM specifications [2].

INTERNATIONAL CLASSIFICATION OF COALS

The Coal Committee of the United Nations Economic Commission for Europe, established in 1949 to develop an international system for classifying coal, after considerable study developed a system which has been approved by the International Organization for Standardization (ISO) [6, 7, 8]. On tests for coking properties of coals, reference should be made to copies of the documents comprising the ISO standards, as well as to the necessary ASTM standards. The details of this international classification are too extensive for inclusion in this manual, but attention will be focused on the salient features. Figure 2 illustrates the classification chart for hard coals.

The ISO classification divides coal into two principal categories. The first is designated "hard" coal and defined as coal having a gross calorific value of more than 10,260 Btu/lb on a moist, ash-free basis. This grouping includes all coals above Subbituminous B in the ASTM classification. The other ISO category includes the ASTM groups designated as Subbituminous B and C, and Lignite A and B. The term *type* in the international classification system is equivalent to *rank* in the American system. These terms refer to the degree of metamorphism, or progressive alteration from lignite to anthracite. The ISO *class* approximates the ASTM *group*.

Coals are first classified according to their volatile-matter content on a dry, ash-free basis, as follows:

Class No.	*Percent Volatile Matter (Dry, Ash-Free Basis)*
1A	3-6.5
1B	6.5-10
2	10-14
3	14-20
4	20-28
5	28-33
6-9	>33

For coals containing more than 33 percent volatile matter:

Class No.	*Gross Btu Value (Moist, Ash-Free Basis)*	*Approximate Volatile Matter Limits (Dry, Ash-Free Basis)*
6	>13,950	33-41
7	12,960-13,950	33-44
8	10,980-12,960	35-50
9	10,260-10,980	42-50

For lignitic and brown coals, or coals having a gross heating value of less than 10,260 Btu, the coals are classified in accordance with their total moisture content on an ash-free basis. Total moisture is that of freshly mined coal:

Class No.	*Percent Total Moisture (Ash-Free Basis)*
10	<20
11	20-30
12	30-40
13	40-50
14	50-60
15	60-70

ISO classes are further divided into groups on the basis of results of standard tests, which will be described in Chapter 4 of this manual. The nine classes of hard coal are divided into groups according to their caking properties in the following manner:*

Group No.	*Free-Swelling Index*	*Roga Index*
0	0-½	0-5
1	1-2	5-20
2	2½-4	20-45
3	>4	>45

The coal groups are further subdivided according to coking properties, as measured by dilatation (using the Audibert-Arnu method), or by the Gray-King coke type:

Subgroup No.	*Maximum Dilatation*	*Gray-King Coke Type*
0	Non-softening	A
1	Contraction only	B-D
2	0 and less	E-G
3	0-30	G1-G4
4	50-140	G5-G8
5	>140	Above G8

A three-digit code number is used to express the classification of a coal. The first digit indicates "class," the second indicates "group," and the third digit indicates "subgroup." As an example, Code No. 634 refers to a coal fitting into Class 6, Group 3, and Subgroup 4.

The classes for subbituminous coals and lignites having gross heating values of less than 10,260 Btu (or Classes 10 to 15) are divided into group numbers based upon tar yield, as shown in Table 2.

Class 10 to 15 coals are designated by a four-digit code number, the first two digits representing class and the last two representing the group. For example, Code No. 1210 indicates a coal that is in Class 12 and assigned to Group 10.

COMPARISON OF ASTM AND ISO CLASSIFICATIONS

The ASTM and ISO classifications compare fairly well (Figure 3). The international system is similar to the ASTM system in that the coals are first

*The terms *caking* and *coking* are frequently used interchangeably, but *caking* generally refers to weakly coking coals. Caking particles adhere, or agglomerate, in contrast with non-caking particles, which form a dust when heated to coking temperatures. In the international system, caking properties are a measure of behavior when coal is heated rapidly, as in the free-swelling test or the Roga test.

Table 2. International Classification of Coals with Gross Heating Values Less Than 10,260 Btu

Group No.	*Group Parameter Tar Yield (Dry, Ash-Free, Percent)*	*Code No.*					
40	>25	1040	1140	1240	1340	1440	1540
30	20-25	1030	1130	1230	1330	1430	1530
20	15-20	1020	1120	1220	1320	1420	1520
10	10-15	1010	1110	1210	1310	1410	1510
00	<10	1000	1100	1200	1300	1400	1500
	Class No.	10	11	12	13	14	15
Class Parameter, Total Moisture, Ash-Free, Percent		<20	21-30	30-40	40-50	50-60	60-70

Source: Reference 6

separated into classes according to volatile matter and calorific value. ASTM uses names instead of numbers, however; and the ASTM system is on a mineral-matter-free basis, while the international system is on an ash-free basis. An internationally acceptable basis for determining mineral matter in coal is not yet available, but methods are now being considered.

In the Subbituminous C coal and lignites, the international system uses total moisture parameters, whereas the ASTM classification is according to calorific values. Good correlation exists, however, between total moisture and calorific value in low-ranked coals.

There is close correlation between the two systems, except where the international system subdivides into narrower categories than ASTM. This is particularly true of the lower ranked coals, where the international system divides lignitic coals into six classes.

Table 3 shows the British standards for coal classification, which are used by the National Coal Board of Great Britain.

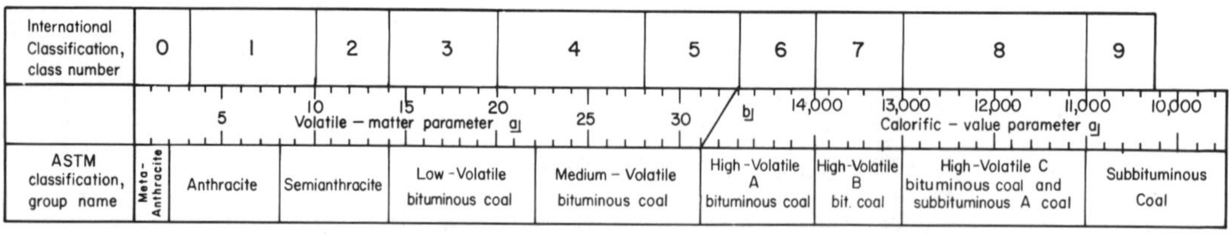

a Parameters in international system are on ash-free basis; in ASTM they are on mineral-matter-free basis.

b No upper limit of calorific value for class 6 and High-Volatile A bituminous coals.

Figure 3. Comparison of the International system (class numbers and boundary lines) with the ASTM system (group names and boundary lines). From Reference 6.

Table 3. The Coal Classification System Used by the National Coal Board (Revision of 1964)

Coals with ash of over 10 percent must be cleaned before analysis for classification to give a maximum yield of coal with ash of 10 percent or less.

Coal Rank Code			*Volatile Matter (d.m.m.f.) (percent)*	*Gray-King Coke Type**	*General Description*
Main Classes	*Class*	*Sub-class*			
100			Under 9.1	A	Anthracites
	101†		Under 6.1	A	
	102†		6.1-9.0	A	
200			9.1-19.5	A-G8	Low-volatile steam coals
	201		9.1-13.5	A-C	Dry steam coals
		201a	9.1-11.5	A-B	
		201b	11.6-13.5	B-C	
	202		13.6-15.0	B-G	Coking steam coals
	203		15.1-17.0	E-G4	
	204		17.1-19.5	G1-G8	
300			19.6-32.0	A-G9 and over	Medium-volatile coals
	301		19.6-32.0	G4 and over	Prime coking coals
		301a	19.6-27.5	G4 and over	
		301b	27.6-32.0	G4 and over	
	302		19.6-32.0	G-G3	Medium-volatile, medium-caking or weakly caking coals
	303		19.6-32.0	A-F	Medium-volatile, weakly caking to non-caking coals
400-900			Over 32.0	A-G9 and over	High-volatile coals
400			Over 32.0	G9 and over	High-volatile, very strongly caking coals
	401		32.1-36.0	G9 and over	
	402		Over 36.0	G9 and over	
500			Over 32.0	G5-G8	High-volatile, strongly caking coals
	501		32.1-36.0	G5-G8	
	502		Over 36.0	G5-G8	

Table 3. The Coal Classification System Used by the National Coal Board–Continued

Coal Rank Code					
Main Classes	*Class*	*Sub-class*	*Volatile Matter (d.m.m.f.) (percent)*	*Gray-King Coke Type**	*General Description*
600			Over 32.0	G1-G4	High-volatile, medium-caking coals
	601		32.1-36.0	G1-G4	
	602		Over 36.0	G1-G4	
700			Over 32.0	E-G	High-volatile, weakly caking coals
	701		32.1-36.0	E-G	
	702		Over 36.0	E-G	
800			Over 32.0	C-D	High-volatile, very weakly caking coals
	801		32.1-36.0	C-D	
	802		Over 36.0	C-D	
900			Over 32.0	A-B	High-volatile, non-caking coals
	901		32.1-36.0	A-B	
	902		Over 36.0	A-B	

Source: National Coal Board (reprinted by permission).

*Coals with volatile matter of under 19.6 percent are classified by using the parameter of volatile matter alone; the Gray-King coke types quoted for these coals indicate the general ranges found in practice, and are not criteria for classification.

†In order to divide anthracites into two classes, it is sometimes convenient to use a hydrogen content of 3.35 percent (d.m.m.f.) instead of a volatile matter of 6.0 percent as the limiting criterion. In the original Coal Survey rank coding system the anthracites were divided into four classes then designated 101, 102, 103 and 104. Although the present division into two classes satisfies most requirements it may sometimes be necessary to recognize more than two classes.

Notes: Coals that have been affected by igneous intrusions ("heat-altered" coals) occur mainly in classes 100, 200 and 300, and when recognized should be distinguished by adding the suffix H to the coal rank code, e.g. 102H, 201bH.

Coals that have been oxidized by weathering may occur in any class, and when recognized should be distinguished by adding the suffix W to the coal rank code, e.g. 801W.

3 Chemical analyses for coking coals

General tests for quality include those normal or conventional analyses that are required for any coal and should be obtained as part of the study of a coal's characteristics. They would include proximate and ultimate analyses and also, the analyses of the ash and the ash fusion temperatures. Determination of the coal's specific gravity should be made. The specific energy (calorific value) of the coal should be obtained. One should also include the free-swelling index (crucible-swelling index) and the Hardgrove grindability index. The latter is a physical test, but is frequently included along with chemical analyses.

Since most coals (especially those destined for coking or metallurgical uses) require cleaning, or "washing," washability (float and sink) tests should be conducted at various relative densities, and an economic recovery of "clean coal" should be determined. Obviously, the various analyses should be made on this "clean-coal" basis, as this best represents the product for end use.

It should be emphasized that no analysis is of value unless the sample being analyzed truly represents the coal being examined. Too often, erroneous results from improper sampling techniques cause disastrous evaluations and conclusions from otherwise accurate chemical or physical tests.

PROXIMATE ANALYSIS

Proximate analysis is the determination, by a prescribed method, of moisture, volatile matter, fixed carbon (by difference), and ash. The details of analytical procedures are given in References 1 and 2, with the ASTM procedure being the one to which reference is generally made.

The term *moisture* can mean the amount of surface or extraneous moisture; the inherent, or natural-bed, moisture; or the combined total moisture. Where the last is given, the analyses are often reported on an "as-received" basis. When given on an "air-dry" basis, the results include the inherent moisture. "Dry" basis is, of course, without moisture. For purposes of designating rank of coal, results are given on a "moisture and ash-free" basis, or they can be given on a "moisture and mineral-matter-free" basis. (See the formula for the latter in Chapter 2.) The percentage of mineral matter in coal is not the same as percent ash, the latter being obtained as the residue from prescribed combustion procedures.

One can see, then, the importance of knowing on what basis the proximate analysis is given. This also affects two other analyses that are generally part of "complete" proximates, namely, the sulfur and calorific values.

With sulfur playing such a vital role in both steaming and coking coals, the sulfur content is almost invariably included in the proximate analysis. It is generally reported as "total" sulfur. Sulfur, however, is in three forms in coal: pyritic, sulfate, and organic. Pyritic sulfur is iron sulfide. Sulfate, usually the result of various sulfate compounds such as gypsum, is generally present in small quantities. Organic sulfur is sulfur chemically bonded in the hydrocarbons of the coal.

Any thorough study of a coal should require the determination of sulfur in all three forms. Considerable effort is being undertaken to lower the sulfur content by various coal washing methods or more esoteric treatment. Since only the pyritic form is reduced by current commercial methods, it is important to know the pyritic content and its dissemination in the coal particles. The methods for determining sulfur content are described in References 1 and 2.

Although not normally required in determining the cokability of coals, it is certainly desirable to know the specific energy or calorific value of a coal. In the U.S.A., this is given in Btu/lb of coal. In some other areas, it is reported as calories/kilogram (cal/kg). To change Btus to cal/kg, multiply by 0.2520. The newer SI metric system is to report the heating value in megajoules/kilogram, or 429.923 Btus = 1 MJ/kg. The heating value can also be reported in calories per gram or joules per gram.

As to the method for determining heating value, ASTM uses the adiabatic calorimeter. Results are reported as "gross heating value," and this is what is meant unless otherwise reported. There is also a "net heating value." This is calculated from the gross heating value by deducting the heat required to remove the water formed by combustion.

Proximate analysis for coke is naturally required also. The exact procedures are somewhat different due to the physical nature of coke, but one ends up with a similar type of analysis. These procedures are fully described by the ASTM in Reference 2.

PROXIMATE ANALYSIS EVALUATION FOR COKING COAL

As previously described, the proximate analysis is the basic determination for any coal, including coking coals. The elements analyzed have important bearings on a coal's "cokability." One determines the rank of a coal by the proximate analysis (ASTM), and rank determines whether a coal falls into the coking-coal category.

Volatile matter is a most important element, since this helps to classify the coking coals and largely determines the fixed carbon limits. Since carbon is essential for reduction in the blast furnaces and volatile matter is largely removed in the coke ovens, a low-volatile coal will give a higher percentage of coke and, therefore, more carbon per ton of coal.

However, coals below 14-percent volatile matter (dry, mineral-matter-free basis) will not coke. Also, at the other end, coals of extremely high volatile matter either will not coke or will make a poor quality coke, and will be low in fixed carbon content. Although this is arbitrary, possibly 38-percent volatile matter could be considered the upper limit.

These limits are only part of the characteristics of coking coals, as described in this manual, but they can help to rule out many coals. Incidentally, the ideal volatile content for making a good coke is in the range of 28 to 30 percent for standard, byproduct-slot-type ovens. As few coals are within this ideal range of volatile matter, coal blending comes into the picture. Although the above range is not binding, since local availability may be the end determining factor, many coke oven operators will aim for a final mix having a volatile content in this range.

The ash content is of vital importance, as this likewise determines the "effective carbon" available. The trend is naturally toward requiring lower ash coals for numerous reasons. Many U.S.A. coking coals are in the range of 5 to 8 percent. Worldwide, the range is much higher, going in some cases to more than 12 percent (dry basis).

An increase in ash increases the coke rate and slag volume in a blast furnace. Although the fluxing ability of some coal ash enters into the reactions, it is generally considered that a 1-percent increase in coke ash reduces blast furnace productivity from 2 to 3 percent. The actual economics vary, but some contracts provide a penalty of $1.00 to $2.00 (U.S.) for a 1-percent ash increase over the contract specifications [9, 10, 11, 12].

Sulfur in coal enters the coke at a rate of about 80 to 85 percent of the coal sulfur content and is even more detrimental to blast furnace operation than the ash level. It has been estimated by some that a pound of sulfur will decrease blast furnace capacity by 25 percent more than a pound of ash. This is due to the increase in slag volume required to slag out the sulfur to keep it out of the iron. Figure 4 illustrates the permissible sulfur and effect on the coke rate.

The upper limit for sulfur in coal for coking purposes is generally placed at 1 percent; however, some higher sulfur coals can be used if they are blended with low-sulfur coals to reduce the average. Some authorities state that an increase of 1-percent sulfur could increase the coke rate as much as 71 pounds per net ton of hot metal [10, 11].

Some metallurgical coal contracts may place a penalty of $0.60 to $1.00 (U.S.) for each 0.10-percent increase in sulfur content over the contract specifications. Obviously, with the economic values placed on these two deleterious ingredients in the coal, the levels of ash and sulfur must be carefully considered.

Metallurgical coal buyers have generally set limitations on the moisture content of the coal. Although this varies with local conditions and transportation distances, the preference is for a maximum range of 6- to 8-percent surface moisture. Since it is inert as far as coke-making is concerned, requires heat to evaporate, and lowers the bulk density of the coke oven charge, the lower the moisture content, the better. On the other hand, extremely dry coal presents dust problems in handling; this also must be taken into consideration. Recent practice in some of the newer coke oven installations is to preheat the coal directly ahead of charging. This system is new and expensive, but on the other hand, increases coke oven capacity considerably.

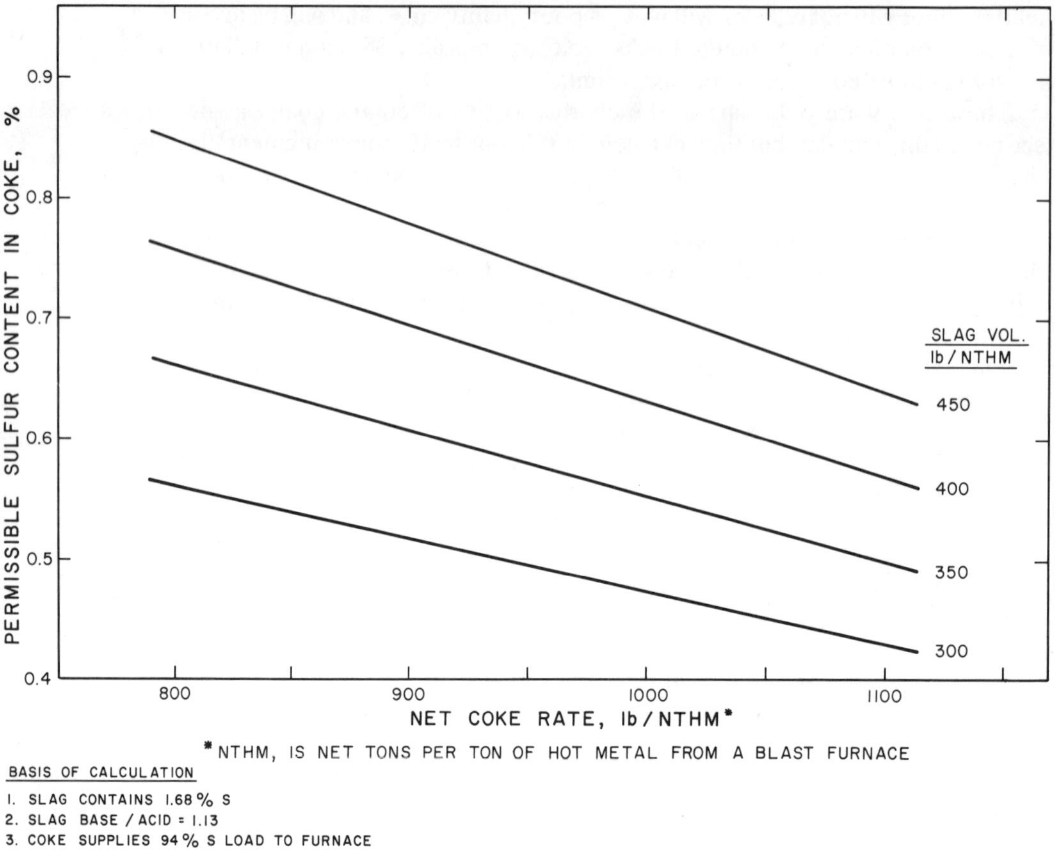

Figure 4. Effect of coke rate and slag volume on permissible coke sulfur. From Reference 10.

Drying at the coal preparation plant (which can be many miles from the point of coke oven utilization) to below 6-percent moisture is costly, since it requires thermal dryers. Wind loss in transit can be substantial, and not the least of the problems are environmental restrictions. However, many contracts for coal, particularly for coals shipped long distances to Japan from North America, hold to this restriction.

ULTIMATE ANALYSIS

The *ultimate analysis* [1, 2] indicates the chemical composition of the coal in terms of hydrogen, carbon, nitrogen, oxygen, and sulfur, on an ash-free basis. The percentage of carbonates is frequently included in this analysis. It may be desirable to present the phosphorus and chlorine contents of the coal in this grouping.

Usually the ultimate analysis is computed either on a dry basis or on a moisture *and* ash-free basis, to avoid inclusion of the hydrogen and oxygen of the water content.

There is a direct correlation between the oxygen content and the quality of coke produced. High-oxygen coals produce relatively weak cokes.

Carbon is, of course, the most important element in the coal, and carbon content is what the coke game is all about. Blast furnace people will use the term *effective carbon* required per net ton of hot metal (NTHM), which is the actual available carbon to do the work of reducing iron ore to metal.

Although not strictly part of what is generally termed *ultimate analysis*, the elements of phosphorus and chlorine, when related as a percentage of the coal (as distinct from the mineral analysis of the ash), are often included.

The phosphorus content is extremely important due to its deleterious effect in iron. All of the phosphorus in the blast furnace burden is reduced and enters the hot metal product. The effect of the phosphorus content in coal on the phosphorus content of the metal is illustrated in Figure 5.

The phosphorus content in coal ash is usually very low, less than 0.3 percent; however, some of the newer coal fields may have a much higher content. Although dependent upon the coke rate, ash content, and analysis of the iron ore, the maximum allowable limit for a 900-lb coke rate and 7.0-percent-ash coke would be 1.7 percent of the coke ash, or 0.09 percent of the coal, assuming a phosphorus content in hot metal of 0.073 percent, which is probably average for most steels. Since most coke is higher in ash than the 7.0 percent quoted above, a good safe limit for phosphorus in coal would be 0.05 to 0.06 percent.

Chlorine, which enters the coal in the form of chlorides, is generally quite low for most coals; there are exceptions, however, and any new coal should be analyzed for this element. The negative effect of chlorine is in the coke ovens themselves, due to its action on coke oven refractories. Most coals are less than 0.10-percent chlorine. No upper limit has been set, but any coal with high chlorides should be viewed suspiciously.

When placing phosphorus, chlorine, and carbonates under the elements in the ultimate analysis, it must be remembered to correct the percent oxygen, as this is generally determined by "difference."

MINERAL ANALYSIS OF ASH

The mineral analysis of the coal ash is of vital importance for steam coal. It is often overlooked in metallurgical coal analyses, but it should be included and more importance is being attached to its content. It is usually determined in the form of oxides; some can have a deleterious effect, while others may aid in the fluxing of the ore.

The most common oxides determined are:

- Silicon as SiO_2
- Aluminum as Al_2O_3
- Iron as Fe_2O_3
- Titanium as TiO_2
- Manganese as Mn_3O_4
- Calcium as CaO
- Magnesium as MgO
- Sodium as Na_2O
- Potassium as K_2O
- Sulfur as SO_3
- Phosphorus as P_2O_5

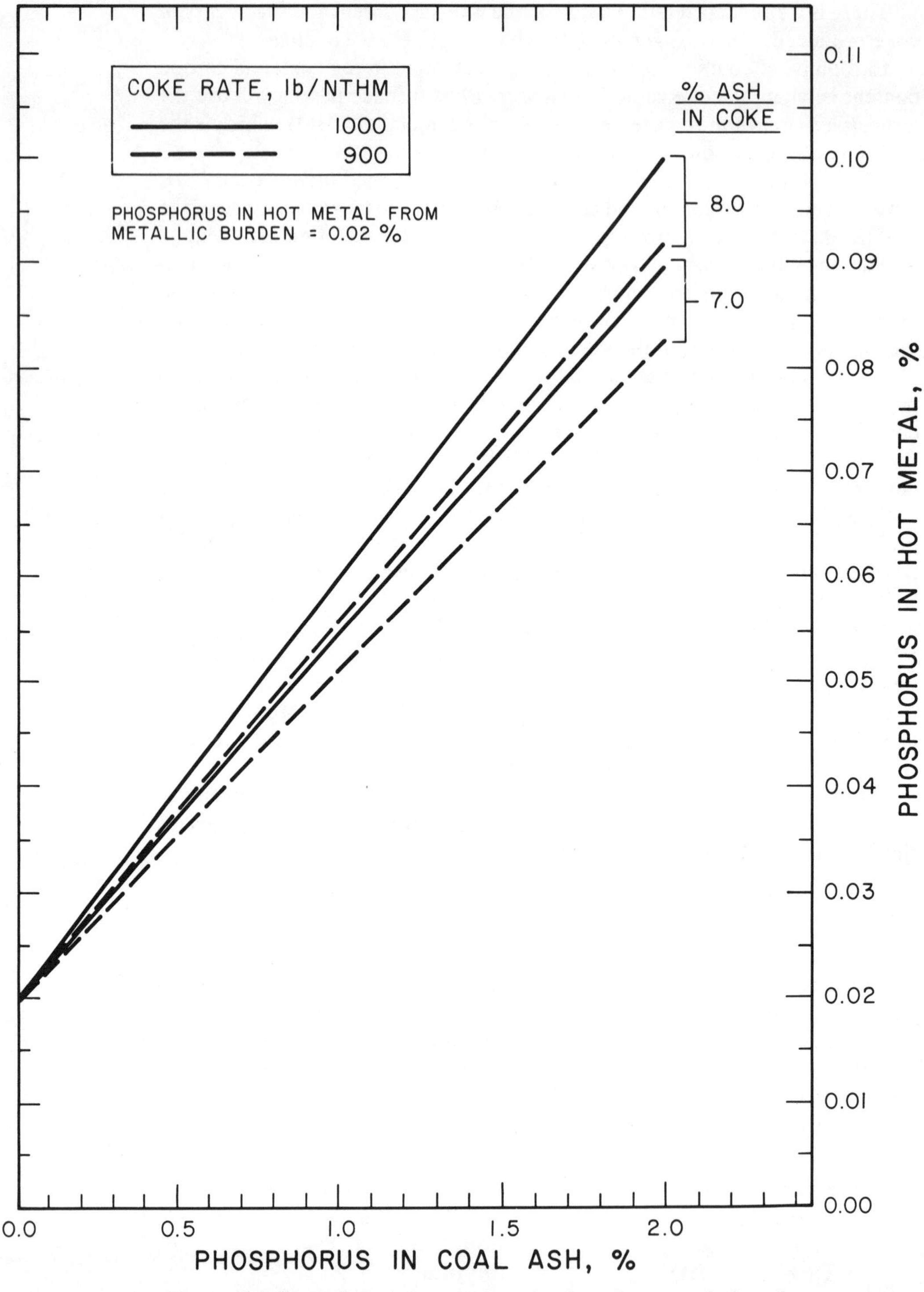

Figure 5. Effect of phosphorus in coal ash on hot-metal phosphorus content. From Reference 10.

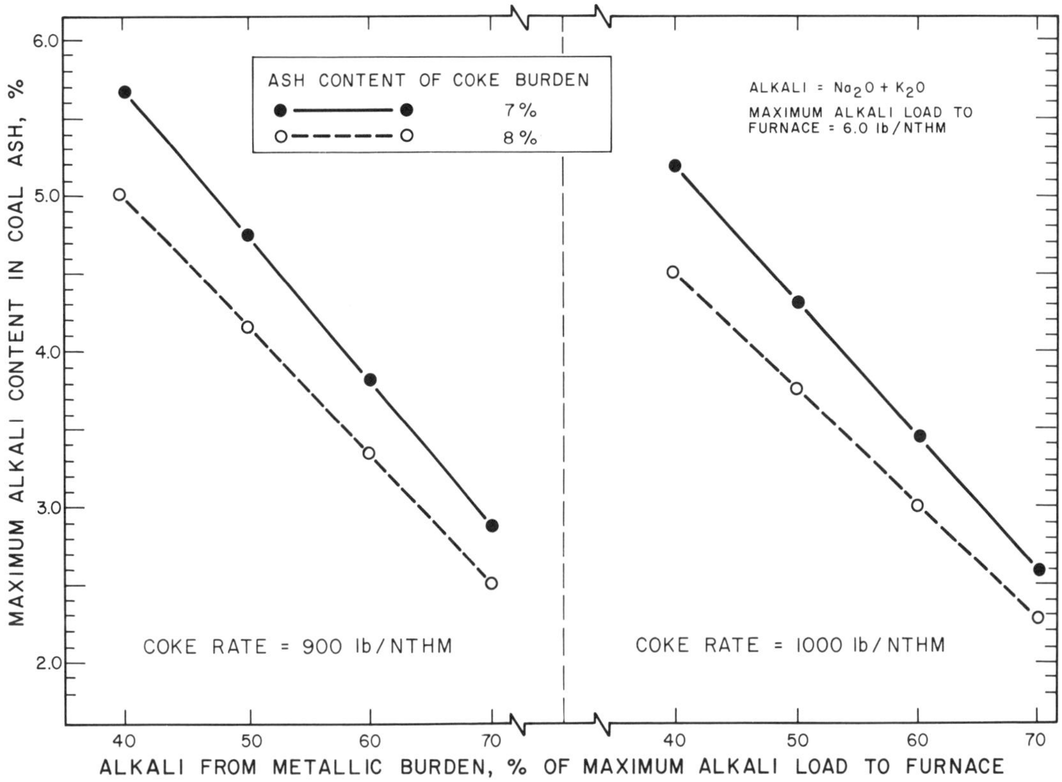

Figure 6. Effect of blast furnace burden on maximum allowable alkali content in coking coal blend. From Reference 10.

It is necessary to determine the base/acid ratio, the slag viscosity, the alkalies as Na_2O, and the water-soluble alkalies as Na_2O and K_2O (dry coal basis).

The alkalies have been found to have a serious effect in blast furnaces, causing decreased burden permeability, lower productivity, and high coke rates. Although present in all raw materials, generally about half of the input comes from coke ash. Figure 6 shows the effect of alkalies on blast furnace burden at two different coke rates [10]. Note that alkalies shown in the figure are the percentage in the coal ash and not the coal itself, and that the permissible quantity varies with the alkali from the metallic burden.

FUSION ANALYSIS OF ASH

The *fusion analysis of ash,* important for steam coal use, is often eliminated from metallurgical coal analyses, but nevertheless should be included as part of the total picture. This analysis is important in determining the melting points of coke ash in the blast furnace, as it can slow up or enhance the slag formation. Standard fusion analyses [1, 2] are determined in both oxidizing and reducing atmospheres, although in the case of blast furnaces, the atmosphere is reducing [14].

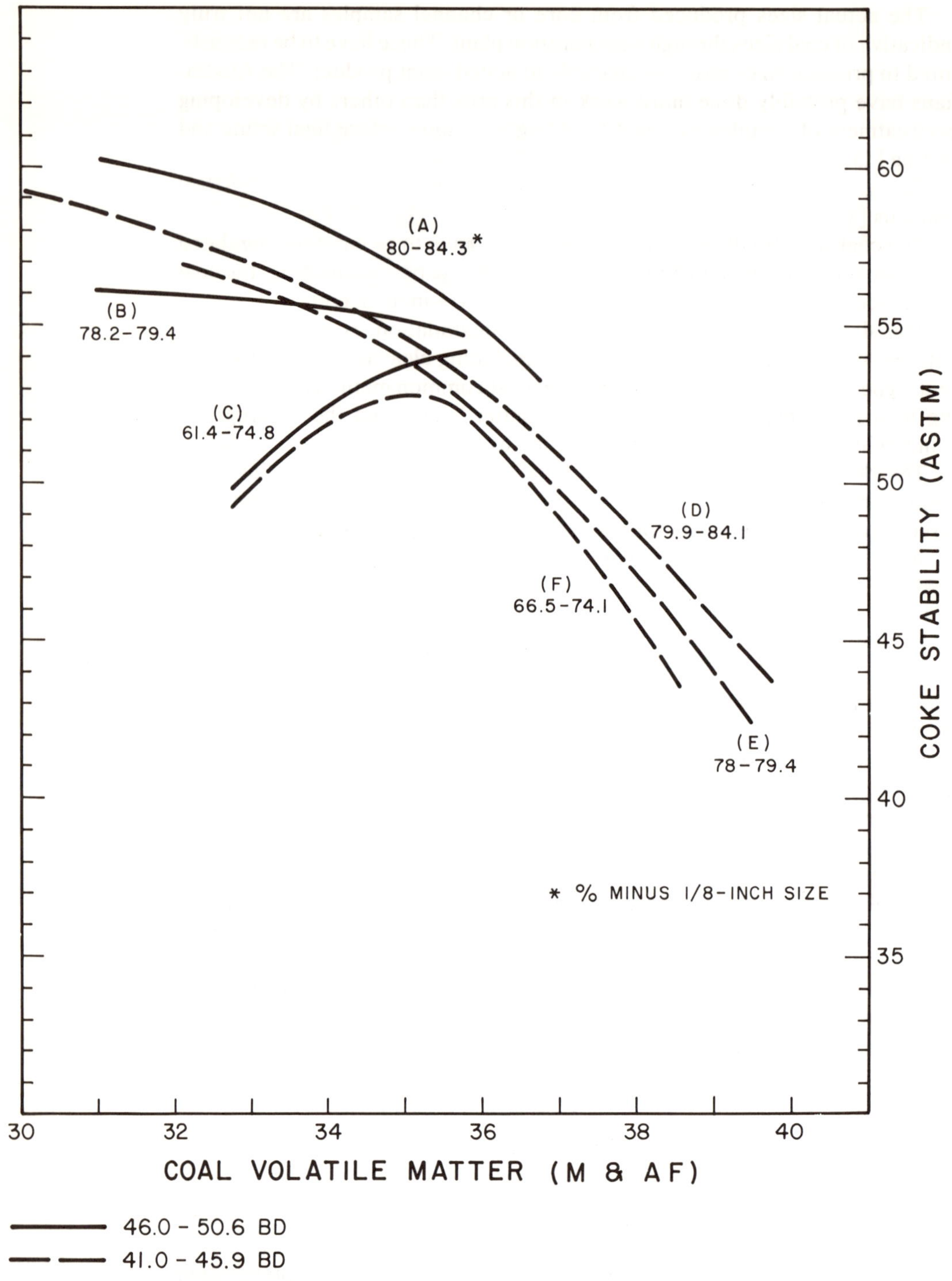

Figure 7. Coke stability versus coal volatile matter for various coal bulk densities and pulverization levels for Hardgrove grindability indices from 58 to 69. From Reference 16.

can be achieved by coal washing but also make it possible to determine which density fractions can produce the most desirable coking characteristics. Such procedures can often show that some coking coal may be obtained from coals otherwise considered "steam grade" or non-coking.

PREFERENTIAL CRUSHING

As will be shown in more detail under discussions of petrographic analysis (Chapter 5), it is known that some sizes (as well as certain density fractions) have better coking characteristics than other sizes, and that size is related to the hardness of these fractions.

Durain lithotypes are much harder than vitrain types of coal, and the former is of poorer coking ability. Sometimes this harder variety of coal can be separated by a combination of crushing and screening from the better coking constituents, resulting in an improved grade of coking coal. Obviously, the economics of preferential operations of this kind must be taken into consideration.

PLASTICITY TESTS

The most common tests for plasticity in both the American and international systems are described below.

Free-Swelling Index (FSI)

The *free-swelling index* (FSI) [2, 4], also called the *crucible-swelling index,* is an important criterion in determining whether a coal will coke. This test is relatively simple and can be used in the field to make a preliminary separation of coking and non-coking coals and to provide some measure of relative coking characteristics. It does not reflect such essential coking characteristics as plasticity and actual coke strength and should be used with caution in evaluating a coal's ultimate coking performance.

Nevertheless, it is valuable as a practical guide in cutting off the grade of coking coal in the seam and in determining oxidized or weathered zones. Although this "cutoff" point will vary with local practice and requirements, generally an FSI of 3½ is considered the lower limit for coking coal potential.

The index measures the degree of free swelling of a 1-gram sample of minus-60-mesh (250 μm) coal heated under certain specific conditions in a translucent silica crucible with a ring-handle pierced lid. The sample is heated to 800° to 820°C for 2½ minutes using a Bunsen burner, or for 1½ minutes in an electric furnace. To get uniform results, it is necessary to follow exactly prescribed instructions, as the volatile content determines the heating time.

The lower the profile of the coke "button" formed, the less the free-swelling power and/or caking power of the coal. If no swelling occurs, the coal is non-agglomerating. The profile of the coke button, if any, is measured against a standard profile chart, illustrated in Figure 8. Obviously, a profile of 8 to 9 is high swelling and, other things being equal, should make a high-quality coking coal.

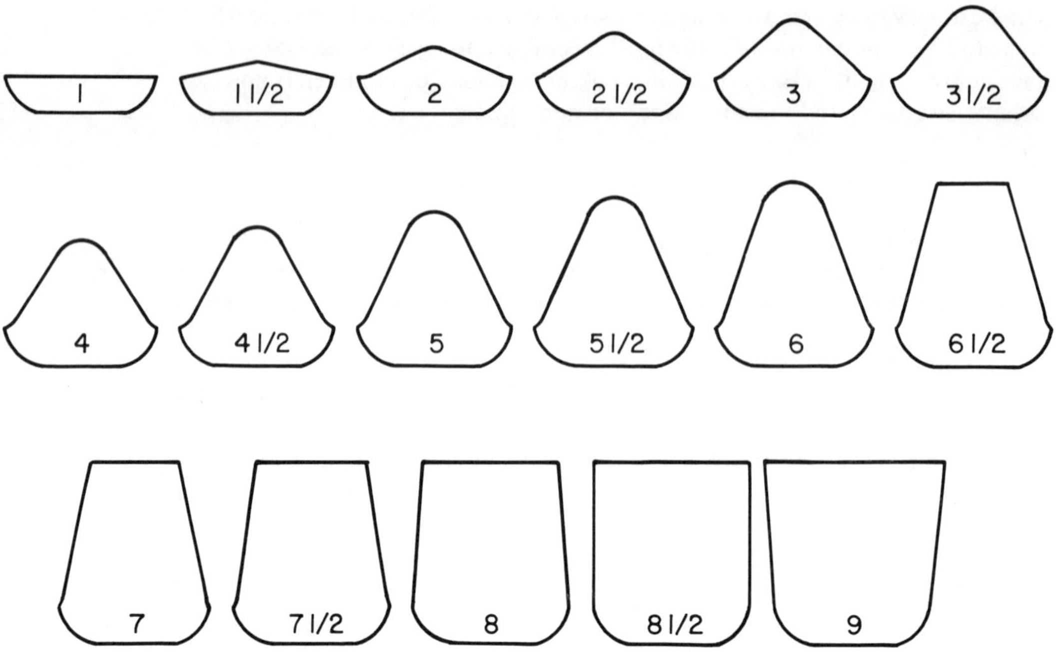

Figure 8. Standard profiles (full scale) and corresponding Free-Swelling Index (FSI) numbers. From ASTM D-720-1972 (reprinted by permission).

The term *agglomerating index* is also used to describe a coal's caking properties. The test was developed by the USBM [1]. It uses the standard volatile matter platinum crucible, and it also measures the appearance and size of the button. It is generally used to determine various degrees of agglomeration and caking rather than free-swelling properties.

The term *agglutinating value* is applied to another USBM test for determining the caking and coking properties of coals [1]. It is somewhat similar to the free-swelling index previously described, but the minus-200-mesh sample is mixed with a sized silicon carbide before being placed in a porcelain crucible and heated to 950°C. Instead of a "profile" evaluation, the resulting sample is given a crushing-strength test, using a compression machine (see ISO Roga test), and is indexed accordingly.

Gieseler Plastometer

Since coking coals are coals which, when heated in an absence of air to sufficiently high temperatures, pass through successive transient stages (first softening, then swelling, then becoming plastic or semiplastic, and finally solidifying into coke), it is apparent that a device for measuring these changes would be of considerable value in evaluating the coking characteristics of a coal or blend of coals.

It is essential that coals melt and become fused into coke. This permits the fluid constituents to combine with the non-melting fractions to form a strong coke. (It can be likened to the blending and fusing of cement and aggregate in making concrete.) Some coals may have too little plasticity; others may be too fluid. In either case, a combination, or blending, of these coals is required. In fact, very few coals are ideal in themselves to make a high-grade blast furnace coke. A knowledge of a coal's plasticity enables one to arrive at the proper blend of various coals.

The Gieseler plastometer is described by the ASTM, currently as D-2639-74. The original testing equipment was manually operated (D-1812-69), and some laboratories still use this equipment. Much greater accuracy is obtained with the constant-torque automatic apparatus, developed in conjunction with Commercial Testing & Engineering Co. of Chicago (CT&E). The Japanese Yoshida automatic apparatus is similar. The details of the apparatus and testing procedures are fully described by the ASTM in Reference 2.

The plastometer measures the plastic properties of coals by the use of a constantly applied torque on a stirrer placed on a crucible into which the coal is charged. The crucible is immersed in a bath, and the temperature is increased uniformly.

The stirrer measures the resistance of the coal particles in the receptacle. It is attached with a hysteresis brake, adjusted until a 40-gram weight is held in suspension. An electric-heated furnace with automatic controls is provided, so that a heating rate of 3.0° ± 0.1°C per minute, on an overall basis, with no more than 3.0° ± 1.0°C for any given minute can be maintained over a temperature range from 300° to 550°C.

As the coal is heated under prescribed conditions, it begins to soften, and the torque weight causes the stirrer to move in the sample. The rate of movement is measured by a pointer-and-dial arrangement. As the coal becomes softer and more fluid, the pointer moves more rapidly up to some maximum, after which, as the temperature continues to rise, the fluidity becomes less and the stirrer moves less until the coal solidifies into coke and the stirrer no longer moves.

The dial divisions are recorded in divisions per minute on cross-section paper against the temperature ranges of plasticity. The *initial softening temperatures* are taken as the movement of 0.5 dial divisions per minute (DDPM). The *maximum fluid temperature* is that at which the dial pointer reaches the maximum rate of movement. The *solidification temperature* is the temperature at which the dial pointer stops. *Maximum fluidity* is the maximum dial pointer movement in dial divisions per minute.

The automated equipment provides for much greater control and accuracy. Temperature readings and dial rates and time are reported on a meter printout.

In addition to the fluidity measurements, the temperature ranges of plasticity and the actual temperatures during this period are equally important, as these have to be taken into account to ensure the blending of other coals during this period of fluidity and temperature. Obviously, the longer the period of plasticity is, the wider the potential for blending with other coals will be.

Although there are some similarities between the Gieseler fluidity measurements and the ISO Audibert-Arnu or Ruhr dilatometer tests, particularly in regard to plastic range and temperature, there is no direct correlation be-

Table 4. Plastic Properties of Coals

	Western Coal Blend	*Eastern Coal Blend*
Free-swelling index	4½	8
Agglutinating value	5.3	8.9
Gieseler plastometer		
Initial softening temp., °C	370	347
Fusion temp., °C	424	404
Temp. of maximum fluidity, °C	437	438
Solidification temp., °C	463	485
Maximum fluidity, ddpm*	16	9300
Audibert-Arnu dilatometer		
Contraction, %	35	30
Dilatation, %	6	210
Softening temp., °C	353	356
Temp. of maximum contraction, °C	411	400
Temp. of maximum dilatation, °C	437	467

Source: References 17 and 18.

*Although practice generally is to report Gieseler dial divisions per minute (ddpm) directly, some, particularly in Japan, will record them in logarithms (to the base 10).

tween dial divisions per minute and dilatations. Table 4 shows some comparisons of plastic properties of coals using both the Gieseler and Audibert-Arnu methods [17, 18].

Figure 9 shows the relationship between Gieseler fluidity and volatile matter [20, 21]. Generally, the higher the percentage of volatile matter, the greater the fluidity, but there is a wide variation in the higher volatile coals due to weathering and other factors.

Audibert-Arnu Dilatometer

The Audibert-Arnu dilatometer test [12, 17, 18, 19, App. A] is an International standard (ISO 349) that is used as a measure of determining the swelling and/or plastic properties of coking coals. It is common in Europe and other parts of the world, although it is just recently beginning to be used in the U.S.A. As explained in the previous subsection, although similar to Gieseler in that the plastic temperature ranges are approximately the same, the fluidities (dilatation) measure different coking coal properties.

A pencil made of powdered coal is inserted in a narrow calibrated tube and topped by a steel rod (piston) which slides in the tube as the coal is heated in accordance with prescribed conditions. By making regular readings of the displacement of the piston as a function of the temperature and expressing the displacement observed in percentages of the original length of the pencil, a curve of the type shown in Figure 10 can be plotted.

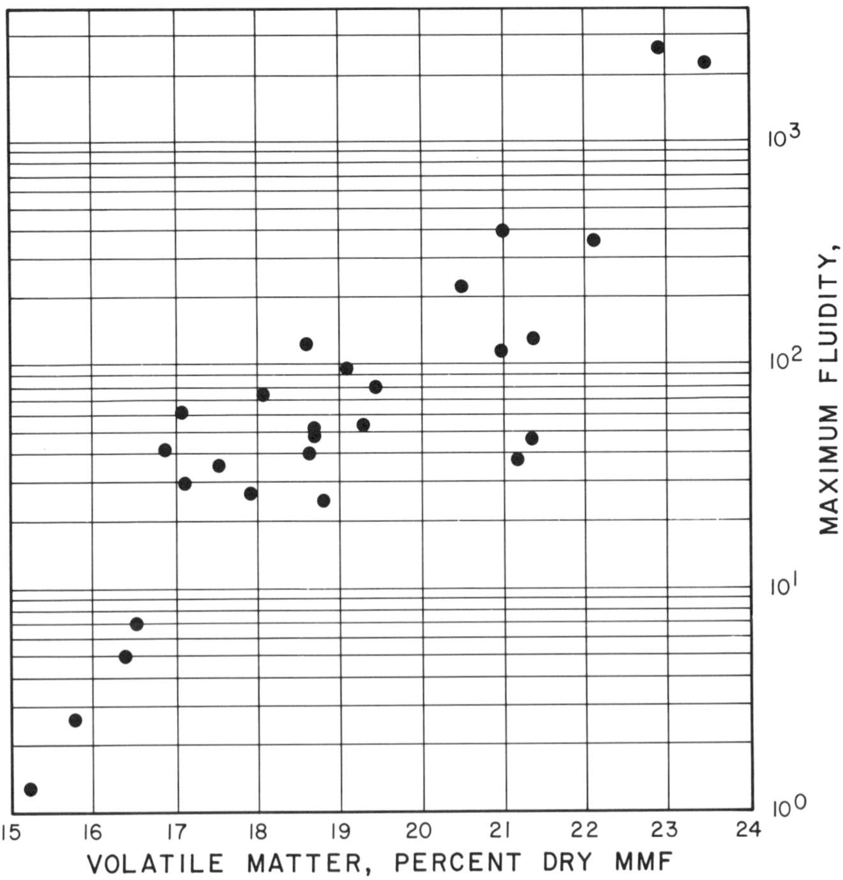

Figure 9. Relationship of Gieseler maximum fluidity of low- and medium-volatile Pocahontas-seam coals to their volatile-matter contents (rank). From Reference 17.

The significant data are as follows:

- T_I = Temperature at which piston has moved 0.5 mm (softening temperature).
- T_{II} = Temperature at which piston reached its lowest point (contraction temperature).
- T_{III} = Temperature at which piston reaches its highest point (temperature of maximum dilatation).
- a = Dilatation, in percent
- b = Contraction, in percent. If, after contraction has taken place, the piston does not return to its original level, the dilatation is calculated from the difference between the final level and the original zero level, but reported as negative (see Figure 10).

It should be mentioned that this test is not designed to indicate pressures exerted on the walls of industrial coke ovens.

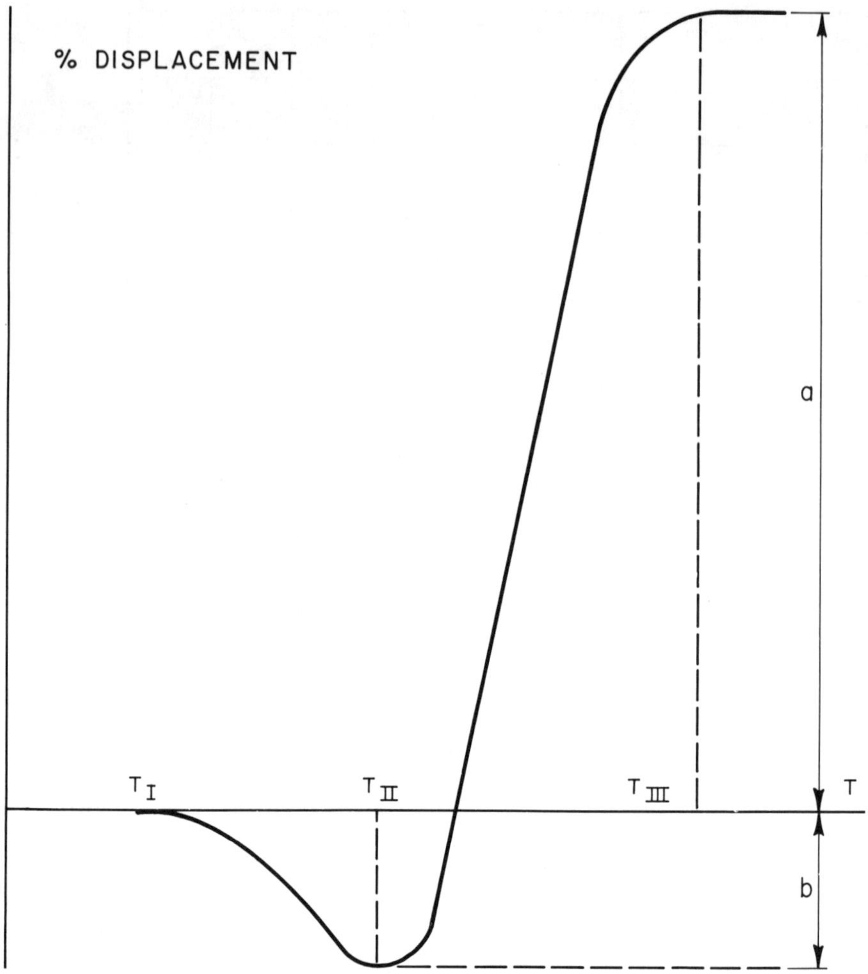

Figure 10. Audibert-Arnu dilatometer curve. From Reference 3.

In view of the fact that the detailed information for this ISO test is not readily available in the U.S.A., the latest ISO Standard (349-1975) is included in this manual as Appendix A.

Although the Audibert-Arnu test is the standard ISO test and is commonly used, some nations prefer the Ruhr dilatometer. This is particularly true in Germany and of the British Standards Institution [4]. The principles are the same, but there is some variation in equipment and procedures. The Ruhr dilatometer is German Standard DIN-51-739 [22] and is given in detail in Appendix B. It is believed by many that the Ruhr method produces a more accurate reproducibility of results and it is used in the determination of the calculated "G" factor in indexing coking coals versus volatile matter.

Some references are given in the literature to the Hoffmann dilatometer test, but it is based upon the same principles as the tests discussed earlier.

Figure 11 shows a relationship developed by the Japanese between Audibert-Arnu dilatometer dilatations and volatile matter, and indicates the relative positions of various coals. The ideal blend is shown in the cross-hatched area [9].

GRAY-KING TEST

The purpose of the Gray-King coke test [3, 6], one of the parameters adopted for the *International Classification of Hard Coals by Type* by the United Nations Economic Commission for Europe, is to assess the caking properties

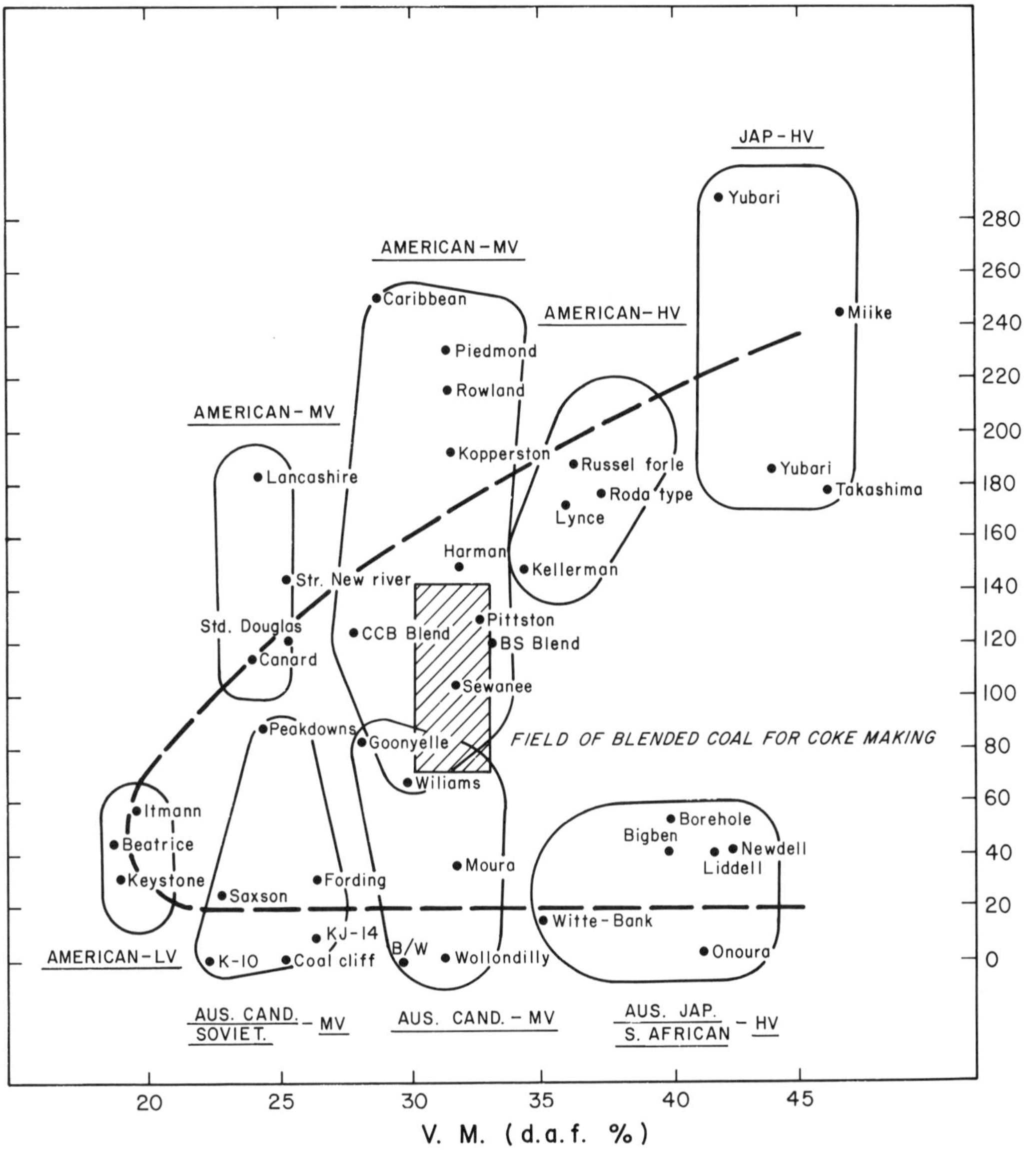

Figure 11. Relationship between dilatometer expansion and volatile matter (V. M.). From Reference 9.

of a coal or blend of coals by carbonization under standard conditions. There is no test comparable to the Gray-King test; it should not be confused with or regarded as an alternative to the Roga test, which also measures caking properties.

The Gray-King test is a laboratory assay method of carbonizing a small, finely ground sample of coal under certain standard conditions up to a final temperature of 600°C. It has been used most frequently in Great Britain.

This test is International standard ISO 502-1974(E) and is described in detail in Appendix C.

The coke residue from the carbonization assay is classified by comparison with a series of standard coke types, grading from A to G. The grades G and G-1 to G-8 represent coking coals, with G-8 being the most strongly coking.

One can make a rough comparison between FSI (crucible-swelling number) and Gray-King as follows:

Crucible-Swelling No.	*Gray-King Coke Type*
1	A to B
1½	C to D
2 to 2½	D to E
3	E to G
4 to 4½	G-2
5 to 5½	G-3
6½ to 7	G-5 or higher
9	G-8

ROGA TEST (ISO)

The Roga test of the ISO [3, 6, 7], originally developed in Poland, is a method for determining the caking power of coal under the standard conditions set forth in its description. It is also used in the ISO standard classification of hard coals. (See Figure 2 of this manual.)

The test is somewhat similar to the USBM agglutinating test, except that instead of silicon carbide, a certain grade of anthracite is mixed with the sample and a tumbler test, rather than a compression test, is made of the product. Anthracite is found to be better than silicon carbide because it will fuse with the bituminous coal tested. See Appendix D for procedures.

A Roga index of 45 indicates a coking coal with an FSI of about 3½ to 4; the stronger coking coals have Roga indices above 45.

SHATTER TEST (ASTM)

The *shatter test* for coke, although not used as frequently as the *tumbler test,* is often referred to and is generally used in determining the strength of foundry coke. It is described in the ASTM procedures [2]. Required per test are 50 pounds of plus-2-in. coke. The coke is classified by measuring its resistance to breakage when dropped from a standard height onto a steel plate. Although the tests are reported as to the percentage retained on 2-, 1½-, 1-, and ½-in.

screens, the results are frequently given only for the plus-and-minus-2-in. coke and may be coupled with the results of the tumbler test.

TUMBLER TEST (ASTM)

The ASTM tumbler test [2] is considered one of the most important tests for determining the strength of coke, especially in terms of the ability of a coke to withstand blast furnace burden. The term *stability factor* is a common measure of the quality of coke, and this factor is used in conjunction with many other tests, including volatile matter and petrographic analyses.

There are several ASTM procedures for coke tumbler tests, since it has been necessary to revise procedures from time to time, basically because modern blast furnaces are using smaller dimensioned coke than previously. The original ASTM specifications were under designation D-294-64 (1972), using 3- by 2-in. (75- by 50-mm) square mesh sieve. ASTM designation D-2490-70 was adopted to run the test on 2- by 1-in. (50- by 25-mm) coke. The latest standard method of tumbler testing for coke is designated D-3402-75.

Tumbler test D-3402-75 is split into Procedures A and B, depending upon coke size. Procedure A for the 3- by 2-in. (75- by 50-mm) size is used when the sample for testing contains a higher weight of material in that size range than in any other 1-in. size range. Procedure B for the 2- by 1-in. (50- by 25-mm) size is used for samples which contain a higher percentage of material in that size range than in any other 1-in. size range.

In Procedure A, 22 ± 0.5 lbs (10 ± 0.25 kg) of dried sample is placed in the drum and rotated at 24 ± 1 rpm underload for 1400 rpms. The product is then screened at 1 in., and the percentage of plus-1-in. (25 mm) remaining is the stability factor. The minus-1-in. is screened at ¼ in. (6.3 mm). The accumulative plus-¼-in. is designated the *hardness factor*.

In Procedure B, size the coke on 2-in. (50-mm), 1½-in. (38.1-mm), and 1-in. (25-mm) screens. Weigh 11 ± 0.25 lbs (5 ± 0.11 kg) each of the dried fractions retained on the 1½-in. (38.1-mm) and the 1½- by 1-in. (38.1- by 25-mm) size. Place both in the drum and rotate at 24 ± 1 rpm for a total of 700 rpms. As in Procedure A, screen the product at 1 in. (25 mm) and ¼ in. (6.3 mm) to determine the stability and hardness factors.

The design and other instructions are shown in the ASTM procedures [2] and are not detailed here. In view of the various changes that have taken place, it is well to know what procedures have been used. Actually, most reported results are based on Procedure A as in D-3402-75 or D-294-64(72). A good coking coal will show a "coke stability" of more than 50 percent plus-1-in.

Figure 12 illustrates the effect on coke stability (strength) of blending various percentages of low-volatile coals with various high-volatile coals.

Micum Tumbler Test

The Micum tumbler test is another drum test to determine coke strength. It is used in Europe and other parts of the world. It is finding increasing use in the U.S.A., in order to meet export coal requirements to Europe.

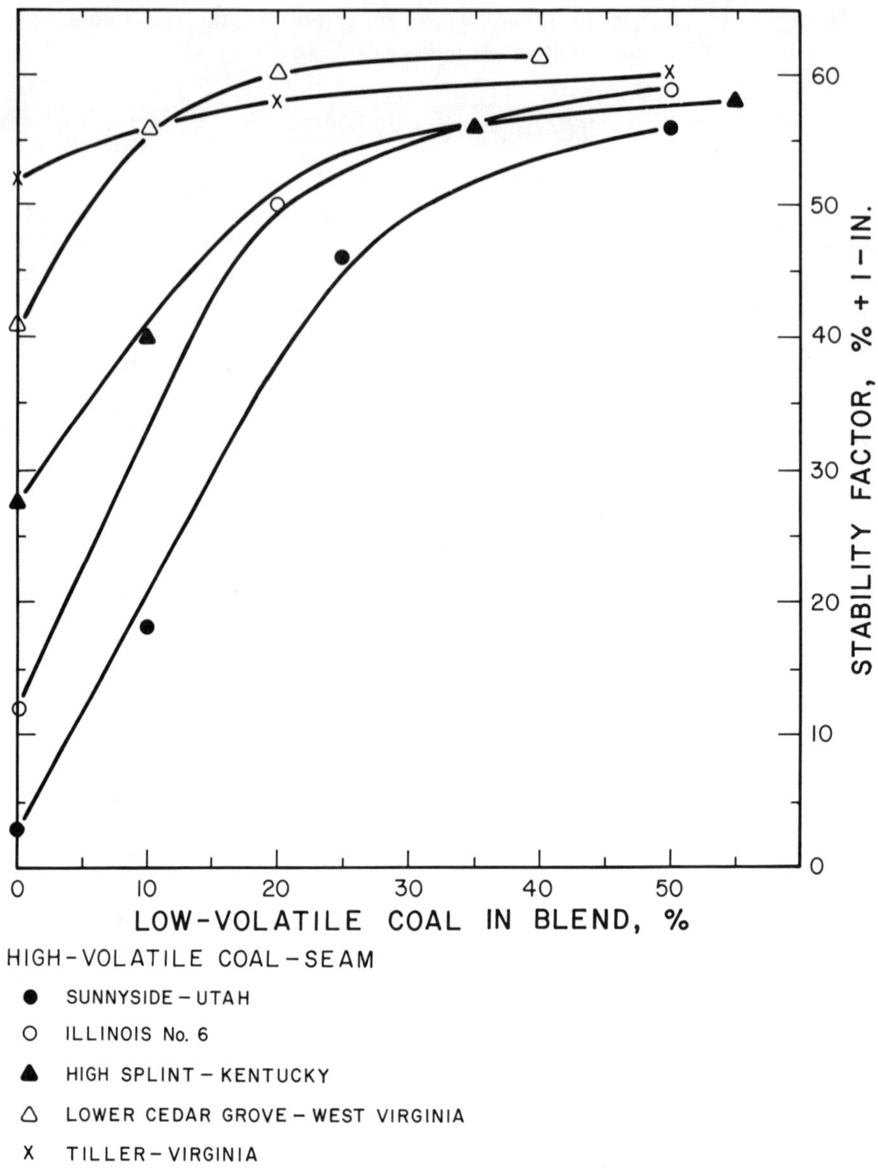

Figure 12. Effect on coke strength of additions of low-volatile coking coal to various high-volatile coals. From Reference 31.

The Micum is an International standard test (ISO TC 27/Secretariat-251), and the procedures are given in Appendix E. The French IRSID (Institut de Recherches de la Siderurgie) is somewhat similar; and, in fact, there is current discussion on combining the two into a modified Micum procedure. The supplement to Appendix E discusses some of the thoughts concerning this matter.

The cylindrical steel drum used in the Micum test is 1000 mm in length and 1000 mm in diameter. Four equally spaced angles are mounted the full length of the drum, which is operated at 25 rpms for 100 revolutions, or 4 minutes.

The sample required is 50 kg with the minus-60-mm screened out on round-hole screens. The products from each test are screened at 40 mm and 10 mm and weighed.

Mechanical strength (Micum M40) is the percentage of the original coke remaining on the 40-mm sieve. *Abrasion index* (Micum M10) is the percentage of the original coke under 10-mm size.

It is somewhat difficult to compare directly the ASTM stability factor and Micum M40, although Montgomery and Verno [13] show some interesting correlations in Figure 13.

Japanese Tumbler Test

The Japanese Industrial Standards (JIS) drum test is used for the standard stability and hardness indices for coal used in Japan, and considerable weight is attached to the results in meriting various coals and blends of coals.

The procedures are described in Appendix F of this manual, which is taken from JIS K 2151 (1960) [5]. The drum is 15 rpms; to determine "stability," as we generally know it, the drum is operated for 2 minutes and designated as drum index (DI) $\frac{30}{15}$. The percentage retained on a 15-mm screen is considered the strength index. Values of 90 or more are classified as strong coke. Some tests are run for 150 rpms and are designated DI $\frac{150}{15}$, as a measure of hardness versus "stability." Frequently, both indices are reported.

Hiroshi Matsuoka [12] shows a relationship between the ASTM stability factor as DI $\frac{30}{15} = 56\ (\text{Stability})^{0.14}$. However, work by Montgomery and Verno [13] shows that where the DI $\frac{30}{15}$ is above 90, there is a substantial range of values in ASTM stability between 50 and 60. This is illustrated in Figure 14.

SOLE-HEATED OVEN FOR EXPANSION-CONTRACTION TESTS

None of the tests previously described have indicated what degree of expansion or contraction would take place in commercial byproduct-slot-type ovens. This is quite important, since expanding coals, such as low-volatile coals, can expand to the point of sticking in the ovens or breaking the oven walls. Maximum coking pressure is generally considered to be about 2 pounds per square inch (psi). It is invariably necessary to blend with low-volatile coals either medium- or high-volatile coals to prevent excessive expansion of coke in the ovens.

Various equipment has been designed to correlate this property. For small samples, a sole-heated oven in which the pressure is measurable is used. About 7 kg of pulverized coal are electrically heated in an apparatus, as shown in Appendix G. It is an official ASTM test, designation D-2014-71, and is fully described in Reference 2.

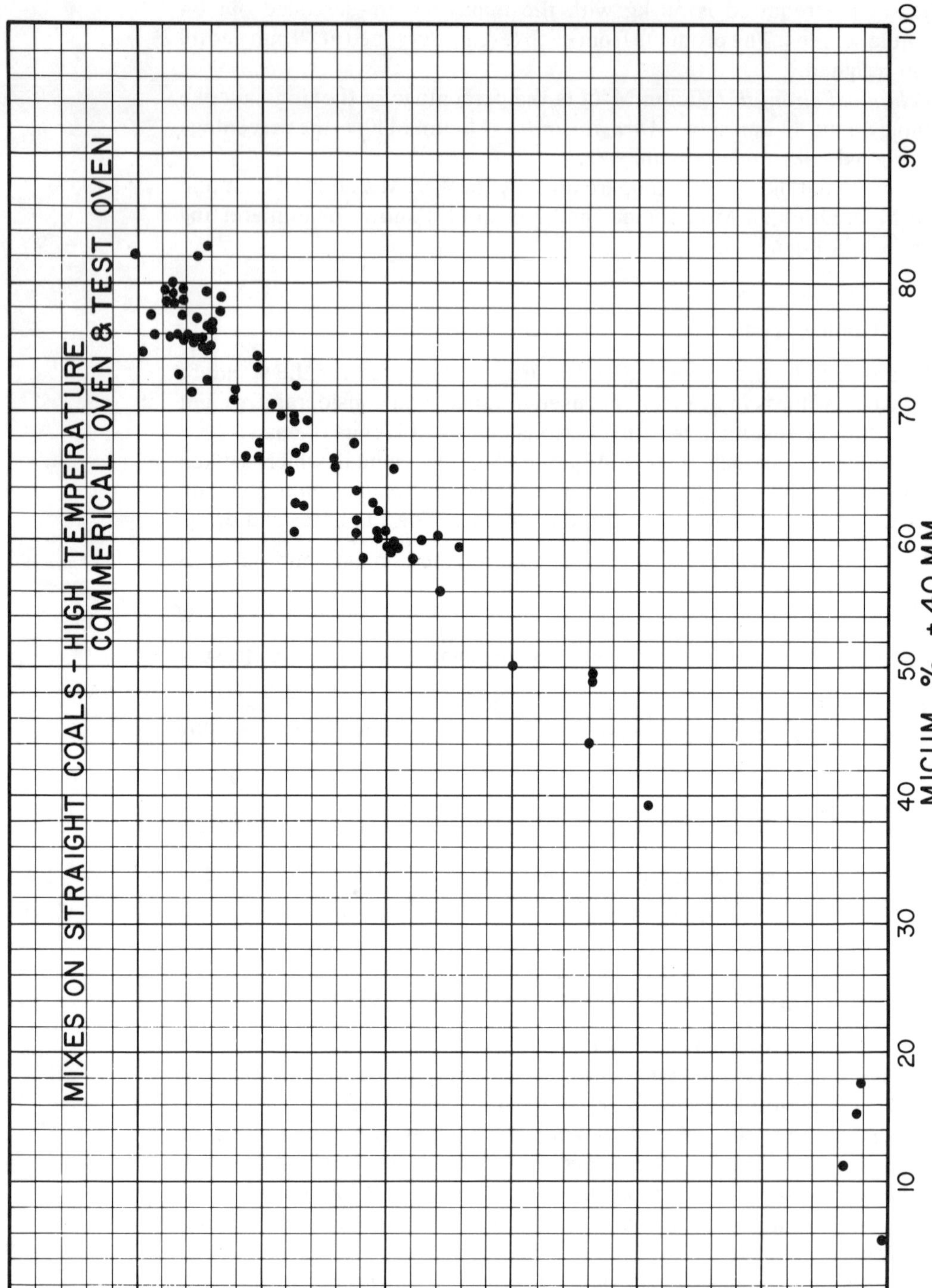

Figure 13. Comparison of American tumbler and European Micum tests. From Reference 13.

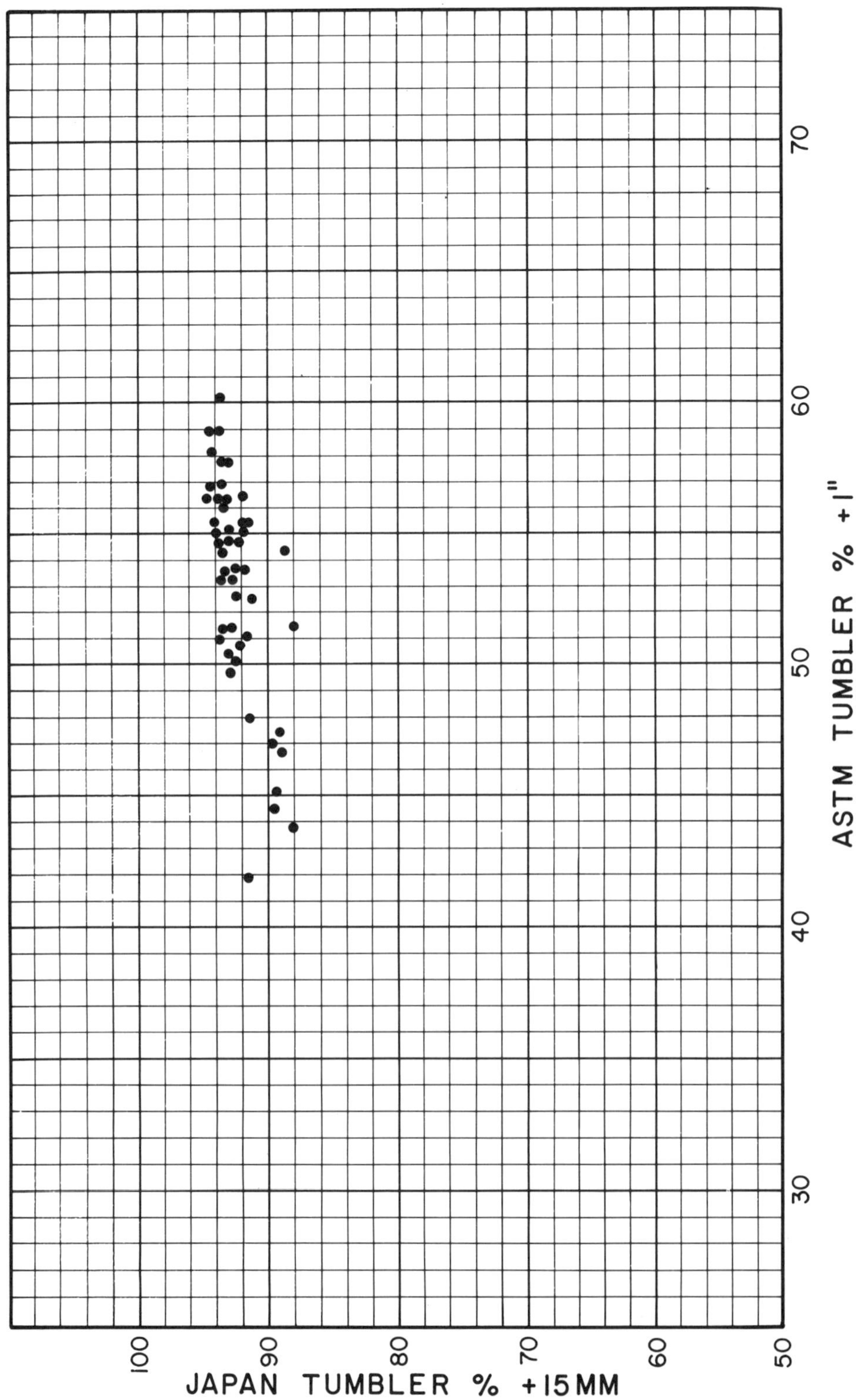

Figure 14. Comparison of Japanese tumbler and ASTM tumbler tests. From Reference 13.

ADJUSTABLE-WIDTH, MOVABLE-WALL COKE OVEN

The sole-heated oven, for small-sample testing, is primarily for measuring expansion-contraction of coal as it is being coked. In addition, there has always been a need for an oven which can simulate large-tonnage commercial coking.

Various research groups have developed test ovens handling anywhere from 400 to 1000 lb of coal charge. With the amount of coke produced from

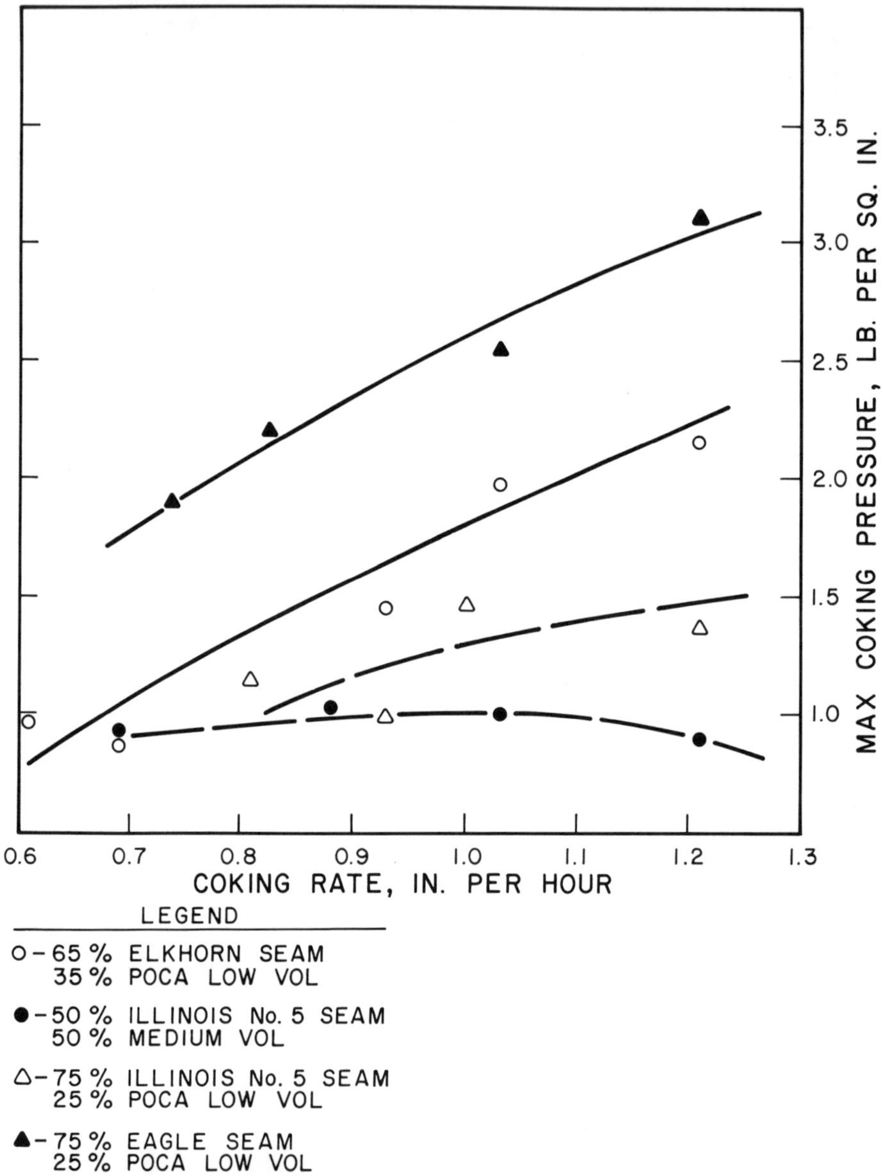

Figure 15. Effect of nominal coking rate on the coking pressure of various coal blends. From Reference 31.

charges of these sizes, one can readily run full-scale coke tests, such as the various tumbler tests. Also, with the movable-wall type, oven pressure can be measured. In addition, heating rates can be changed and oven-width dimensions varied for thickness of charge. Figure 15 illustrates the effect on coking pressure, with various coal blends, of varying the coking rate.

There is no standard oven as such, although they all work on the same basic principles. One of the popular designs, described in a paper by Schoenberger and Kuchta [33], is that developed by the United States Steel Corporation. Drawings of this oven are shown in Appendix H.

In the examination of an unknown coal or blend of coals, it is desirable, although not always possible, to have samples of the coal tested in a movable-wall oven. In fact, most large users of coking coals will insist on checking the coal in these 400- to 1000-lb ovens before final commitment, in spite of the relative accuracy of predictions from the various laboratory tests and analyses.

Various research groups use smaller ovens, and many of the coking properties can be measured by them [34, 35, 36]. Some of these ovens frequently have associated apparatus for measuring and collecting the byproducts, as well as the coke itself.

5 Petrographic analyses to determine coking properties

It has been known for many years that the coking characteristics of a coal are related to its petrographic constituents. Only recently, however, has it been determined that (1) many properties of a coal can be correlated with its petrographic constituents, and (2) it is possible to predict characteristics, such as stability (strength) and hardness, of the coke produced from a coal. As a consequence, petrographic analysis of coal has become a major analytical tool. In fact, some of its proponents use this method to replace many of the tests previously used for evaluating coking coals.

Petrology is the earth science dealing with the study and classification of the rock strata of the earth's crust. Coal petrography is a specialized division dealing with the study of organic sedimentary strata known as coal.

Since rock strata are made up of minerals of inorganic substances, the term *minerals* cannot, by definition, be applied to coal (although there are some minerals in coal). The word *maceral* is used to describe microscopically distinct organic entities in coal. A maceral is to coal as a mineral is to rock [37].

To determine these entities, or macerals, they must be observed under a microscope. Much of the early work was done using thin sections, by transmitted light. This method has long been associated with the work of Thiessen [28], and others. It has considerable value in determining certain basic entities of coal and has helped in determining the botanical origin and rank of coals, but is only of general value for evaluating coking coals.

It was the development of reflected light methods under the microscope that opened the way for detailed determination of coking properties. This is accomplished by reflecting a beam of light off a polished surface of coal. As early as 1919, Marie Stopes [38] introduced the idea of the reflected light technique and proposed the classification of coals in accordance with their banded constituents; the method of classifying coals as lithotypes (into vitrain, clarain, durain, and fusain) has long been used among coal technologists.

As the techniques advanced and by combined use of thin-section transmitted light and reflected light methods, the various kinds of macerals were distinguished more clearly from the lithotypes. The Stopes-Herlen system [38], shown in Table 5, was developed.

Table 5. Stopes-Herlen System of Classification

Lithotype	*Maceral*
Vitrain	Vitrinite 1. Collinite 2. Telinite
Clarain	Variable composition 1. Vitrinite - distinctive 2. Exinite a. Sporinite b. Cutinite c. Alginite d. Resinite 3. Inertinite - less distinctive a. Micrinite - fine-grained b. Sclerotinite c. Semifusinite d. Fusinite
Durain	Rich in 1. Inertinite - distinctive 2. Exinite May be present 3. Vitrinite - less distinctive
Fusain	Fusinite

Source: Reference 38

Note: As can be noticed above, the description of coal using lithotype nomenclature is not very exacting. The further subdividing into macerals, which delineate the various organic plant remains which make up coal, are far more accurate.

Improvements were made in the reflected light method as various international petrographers—Seyler of the United Kingdom; Potonie, Stack, and Hoffman of Germany (who utilized an oil immersion lens); the Russians under Ammosov, *et al.* [29]; and Spackman (U.S.A.) [24]—became involved in the work.

Finally, a system of classification was developed by the International Committee for Coal Petrology [24, 26, 28, 29, 30, 38, 41]. The system, used in many parts of the world, classifies the various macerals as shown in Table 6.

As interest increased and techniques advanced, it became natural to determine the effect of these various entities on coal characteristics, including their effect on coking of coals. Emphasis then was centered on determining the effect of these substances in the carbonization of coal and on delineating these substances quantitatively as well as qualitatively.

Parallel work was done in Russia, Germany, and the U.S.A. in this area [23, 24, 26, 28, 29, 30, 31, 39]. In the U.S.A., the work of Shapiro and Gray [24, 26] and of Harrison [28] did a great deal to advance the use of petrographic analyses to determine the coking characteristics of coal. These men developed a classification for the various entity groups of macerals, based upon the differences in reflectance values (in oil), and then by extensive research in coke test ovens were able to tie down the coking characteristics of the various macerals and even the degree of coking characteristics among the subtypes of macerals. Their system, with some modifications in certain countries, predominates as the method of predicting coking characteristics.

Table 6. Reflected Light Classification of Macerals

Maceral Groups	*Microcomponents*
Vitrinite	Collinite Telinite
Semivitrinite	Semicollinite Semitelinite Mixtinite
Fusinite	Semifusinite Micrinite Fusinite Sclerotinite
Leiptinite (Exinite)	Sporinite Cutinite Resinite
Alginite	Algocollinite Algotelinite
Mineral Impurities	Clay Minerals Iron Sulfides Carbonates Other Minerals

Source: Reference 29

Shapiro and Gray [24, 26, 28] developed a classification of coal entities using reflected light, differentiating between reactive macerals and inert macerals (Table 7). They use a somewhat different nomenclature than shown in Tables 5 and 6, but the meanings are the same and can be correlated. Most countries use the term *vitrinite* to identify what Shapiro and Gray term *vitrinoid*; so generally speaking, the words are synonymous. Also, the term *entity* can be considered synonymous with *maceral*.

To be more specific, however, the classifications ending in "-ite" represent "suite" or "groups"; whereas, more closely related macerals are classified by Shapiro and Gray with the ending "-oids." In the U.S.A. the term *vitrinoids*, etc., is generally employed, while other parts of the world will use *vitrinites*, etc. Strictly speaking, one should follow the glossary established by

Table 7. Classification of Coal Entities (Macerals)

General Categories	*Maceral Groups*	*Maceral Types & Minerals*
Vitrinite suite	Reactive vitrinoids	V0 to V21
	Inert vitrinoids	V22 to V70
	Reactive semifusinoids	SF0 to SF21
Leiptinite suite	Exinoids	E0 to E15
	Resinoids	R0 to R15
Inertinite suite	Micrinoids	M18 to M70
	Fusinoids	F40 to F70
	Inert semifusinoids	SF22 to SF40
Mineral suite	Sulfides	Pyrite, Marcasite
	Carbonates	Calcite, Siderite
	Silicates	Illite, Kaolinite, etc.

Source: Reference 26

Notes: Under vitrinoids (-ites) are also the macerals collinite and telinite. Under exinoids (-ites) are sporinite, cutinite, and alginite. Under resinoids (-ites) are resin and waxes.

Semifusinoids are part "inert" and part "reactive"; in other words, a mixture. Average for Appalachian coals and frequently used elsewhere is that this maceral group is one-third reactive and two-thirds inert, and is so distributed. However, this is not necessarily true with other coals and this variance should be considered in analyses; Australian and western (Rocky Mountain) coals may have 50 percent or more of the semifusinites as reactives.

the International Committee for Coal Petrology [38]. In this text, we prefer the term *vitrinite*, etc.

PETROGRAPHIC ANALYSIS BY THE OIL REFLECTANCE METHOD

The actual procedures for making petrographic analyses are complicated; not only is the proper equipment required, but the work should be done by a competent and skilled petrographer.

Once the equipment is set up, however, the procedures are fairly standard. The latest equipment is designed to be more or less automatic and is tied to an electronic data computer (Ameda system or equivalent).

Obviously, great care must be taken to obtain representative samples and in the preparation of the polished briquets. It is not possible within the confines of this manual to describe the procedure fully. It is essential to refer to ASTM D-2796-72, D-2797-72, D-2798-72, D-2799-72 [2], and any later appendices to these requirements. Another good reference is the Bituminous Coal Research manual titled *Preparation and Polishing of Coal and Coke for Petrographic Analysis* [37].

The coal is crushed to pass through a 20-mesh (850-μm) sieve and mixed with an epoxy-resin binder, then placed in a mold (25 mm in diameter). The resulting briquet is polished on the surface and then examined under carefully prescribed conditions.

Quantitative petrographic analysis is made by a Leitz BMe (or equivalent) microscope with a potential magnification of at least 400. The group macerals are counted by a "point count" system, with four points at 1-mm spacing per field of view, to give up to 2000 points per sample. (The value of computerization is apparent.) The reflectance values are made on a Leitz UAM microscope (or equivalent) with an oil immersion objective at a magnification of at least 500. In measuring the reflectance, a peak transmittance range of 546±5 nanometers (nm) is generally used, although some petrographers use a wave length of 526 nm.

Thus, in addition to measuring the quantity of macerals by type, the petrographer also obtains the reflectance of these types. The report includes the maximum mean reflectance of a specified coal component (usually vitrinite) as percent reflectance in immersion oil. The spread of reflectance is indicated in one of three ways—in a table of individual maximum reflectance values, as a frequency histogram, or as a table of percentages within reflectance classes.

INTERPRETATION OF PETROGRAPHIC ANALYSIS

The principal substances in normal coal are the vitrinites. The main coke-forming substances in coal are the reactive vitrinites, together with exinites and resinites and part of the semifusinites (Table 7). As a rule, these last groupings represent a relatively small part of the coal macerals.

During carbonization, the exinites and the resinites volatilize to a greater degree and show low reflectance values, but they aid in the development of vacuoles in the coke structure and increase fluidity of associated vitrinites. Micrinites commonly occur as small particles in attrital coal and behave as relatively inert substances during carbonization. Fusinites also are relatively inert during carbonization. Minerals, of course, are inert.

One notes from Table 7 that not all vitrinitic macerals are reactive in coke-making. For example, high-ranked coals (such as semianthracite or anthracite), which may be high in vitrinite, will not coke and can be classified as inert, in terms of becoming fluid and volatilizing into a coke-making bonding agent.

To determine the cutoff point microscopically between the reactive macerals and the inert macerals, it has been found that reactives have relatively low reflectance values and inerts have high reflectance values (see oil reflectance described in an earlier section).

There is a wide variation in the reflectance values of macerals, not only of the vitrinites themselves but of other entities as well. The petrographer, after first determining the kinds of macerals inherent in the coal (i.e., vitrinites, fusinites, etc.), must also determine the percent of reflectance of the various grades of these entities. This is done by measuring the percent of light reflected from the polished coal surface of the different macerals. This reflectance is termed Ro and is measured in percentages of 0.1 increments.

It has been found that the percent reflectance for reactive macerals will range generally from 0.5 to 2.0 percent. This will vary according to the rank of the coal, i.e., with high-volatile coals showing low reflectance values and low-volatile (higher ranked) coals showing high reflectance values. The coking coal range is from 0.5 to 2.0 percent.

Reflectance values below 0.5 percent represent lower ranked high-volatile and subbituminous coals. Coals above 2.0 percent are non-coking semi-anthracitic and anthracitic classes. Anthracite, for example, may have reflectance values of 7.0 to 8.0 percent.

The high ranges of reflectance also apply to inert constituents and inert vitrinites, which from the coking viewpoint are called non-reactive.

To simplify the description of reflectance values of different macerals, the term *type* is used; each type is a 0.1-percent reflectance, with Type 1 being between 0.1- and 0.2-percent reflectance and Type 5 being between 0.5- and 0.6-percent reflectance, for example.

These types, as shown in Table 7, can go up to 80, or 8.0-percent reflectance (R_o). Tests have determined that the coking reactive maximum range is from about Type 5 to Type 20. Since most of the reactives are vitrinite, some charts show the type classification only of this group, although basically it is necessary to distribute the other reactives in proper proportion.

An example of the distribution of types of vitrinites in accordance with their reflectance values is shown in Figure 16. This histogram also classifies the coals in accordance with their volatile matter (dry, ash-free basis). Figure 17 illustrates a similar relationship, but uses reflectance values instead of vitrinite types.

Shapiro and Gray and many others have shown that there is a direct relationship between the types of reactives and the amount of inerts in making coke. Their work and that of others was based upon experimental coke oven tests on coals of different reflectance values (types or classes) in relation to the quantity of inerts or non-reactives in the coal; they were thus able to come up with the relative strength (stability) values of a coal or blend of coals based upon petrographic analyses.

It has been found that reactives in themselves will not make a good coke but require inert material in proper proportion (similar to cement and aggregate). The amount of inerts required will vary with the types or classes of reactives.

It is necessary then to determine the distribution of the reactive types and the amount of inerts required for each type. Figure 18, based upon considerable research, shows the inerts/reactives ratio required for each reactive type to give the greatest coke strength. Each reactive type has an optimum ratio of reactives to inerts. For example, to achieve the greatest coke strength, Type 13 would need four times as much reactives as inerts.

The optimum amount of inerts rarely exists in coals as they occur in nature. From the types-versus-inerts requirements, however, it can be determined how far out of balance a coal may be. A numerical value, called the balance index or *Composition Balance Index* (CBI), is placed upon this coordinate.

If the coal is in balance, the amount of inerts present will be optimum for the quantity of reactives. This ratio will approach a balance index of 1.0. If the

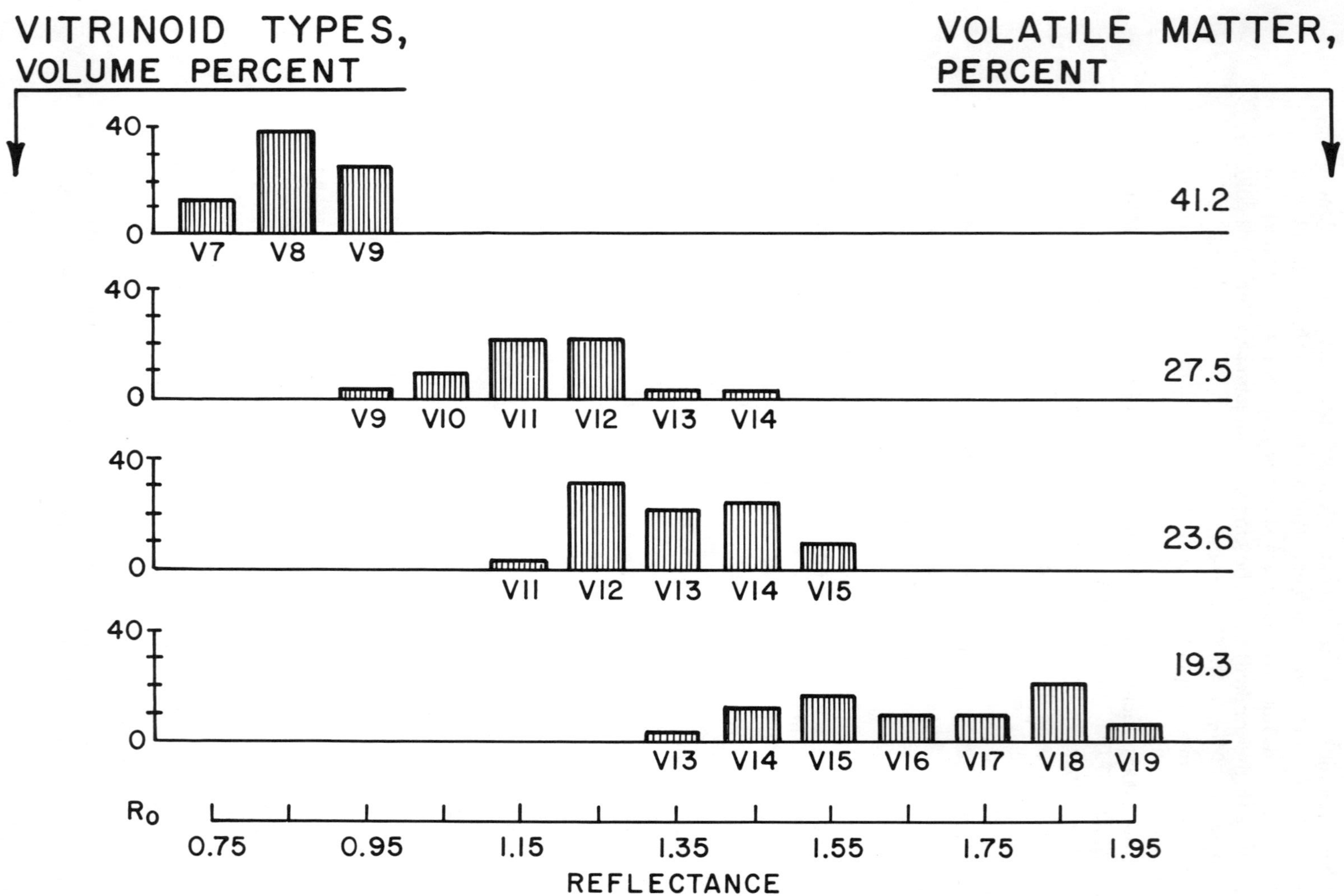

Figure 16. Distribution of vitrinoid (vitrinite) types in coking coals of various ranks. From Reference 36.

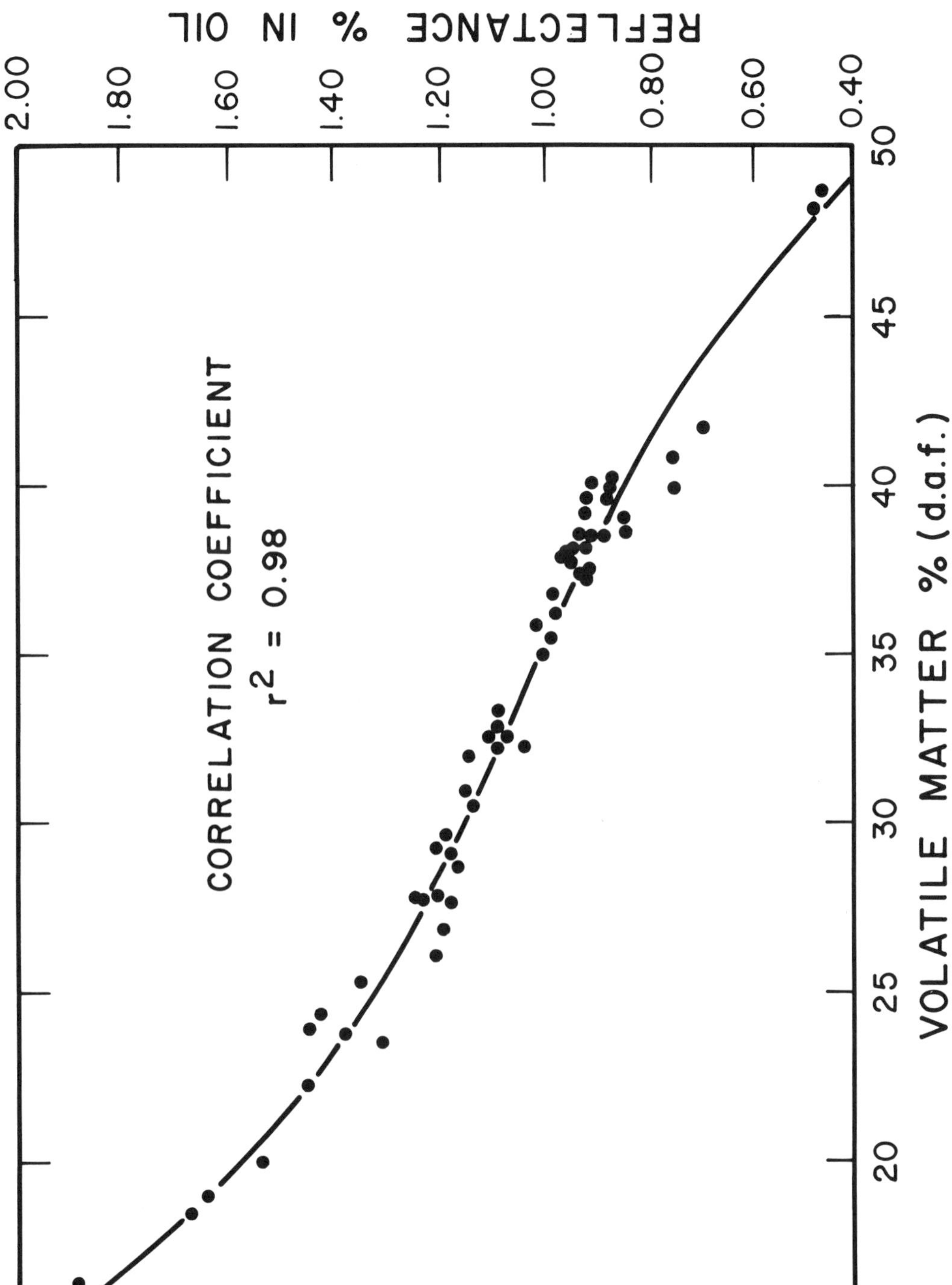

Figure 17. Relationship between reflectance and volatile-matter content of coals. From Reference 36.

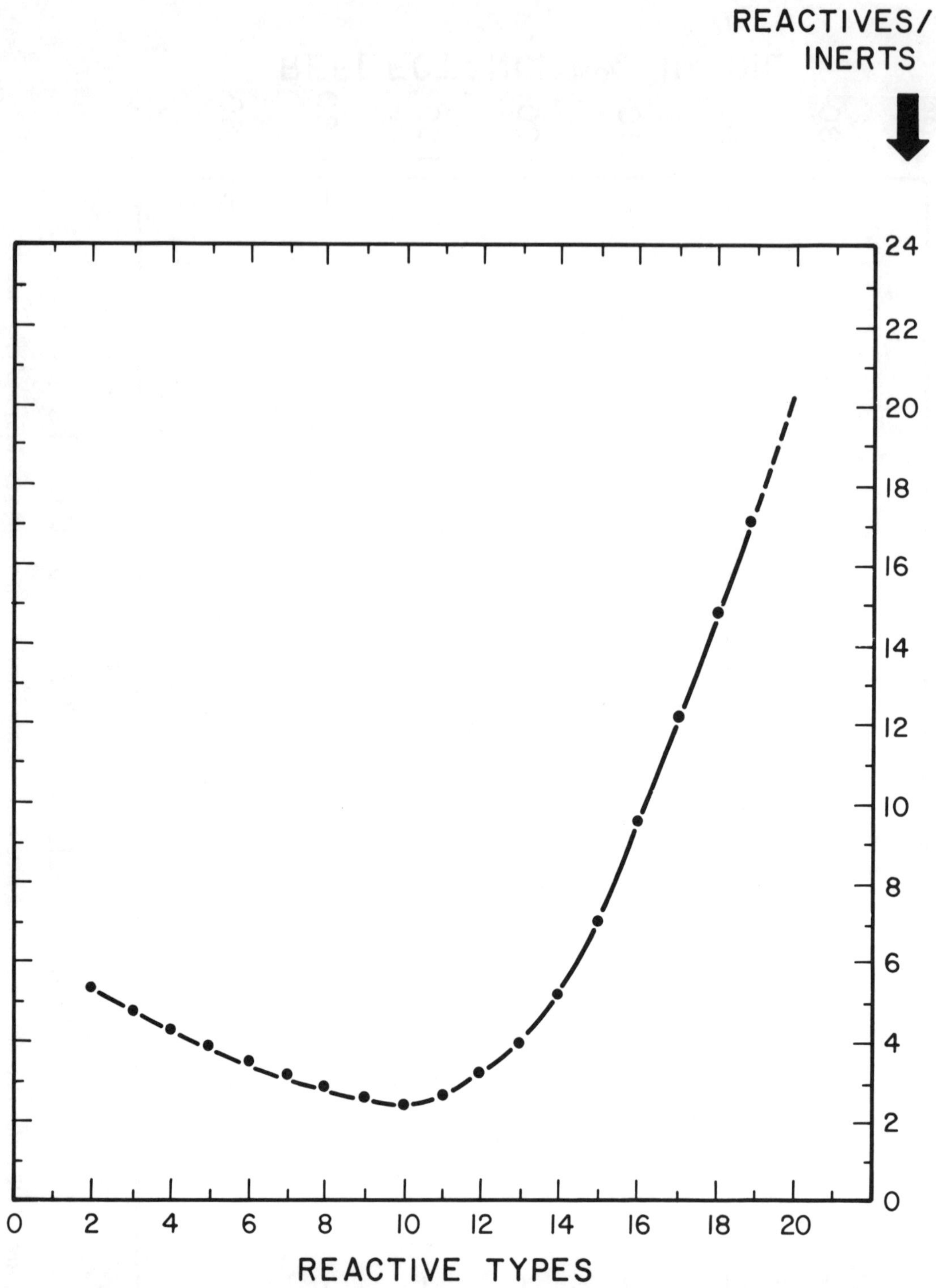

Figure 18. Optimum inerts chart: Optimum ratios of reactive to inert components for reactive maceral types. From Reference 36.

amount of inerts is greater than the amount required, then the index will be greater than 1.0.

Different types show different capacities for assimilation; therefore, ultimate strength differences related to rank or degree of carbonization are experienced.

The graph in Figure 19 is a summation of the amount of inerts and the strength properties of each reactive type. For each type, the curve peaks at the point at which the optimum amount of inerts is present in the mixture. It is apparent that low reflectance types give low relative strengths. Type 5, for instance, has a relative coke strength of 2½. Type 15, with its optimum amount of inerts, has a coke strength of greater than 7.

This is called the *Strength Index* (SI) and is defined as the comparative coke strength of the reactive macerals present in the coal mixture. SI is the second of two parameters required to predict the strength of the coke produced from any coal; the other parameter is the Composition Balance Index (CBI).

The formula for the CBI can be expressed as follows:

$$\text{CBI} = \frac{\text{Inert Percent}}{\frac{P_1}{M_1} + \frac{P_2}{M_2} \ldots \frac{P_n}{M_n}}$$

where:

CBI = Composition Balance Index (also called Inert Index)

$P_1, P_2, \ldots, P_n$ = Percentage reactives in reflectance classes (types) 1, 2, . . ., n

$M_1, M_2, \ldots, M_n$ = "Ideal" ratio of reactives to inerts in reflectance classes 1, 2, . . ., n

The formula for Strength Index (SI) can be expressed as follows:

$$\text{SI (also called Rank Index)} = K_T$$

$$K_T = \frac{(K_1 \times P_1) + (K_2 \times P_2) + \ldots + (K_n \times P_n)}{P_T}$$

where:

$K_1, K_2, \ldots, K_n$ = Strength indices of reactive classes 1, 2, . . ., n

$P_1, P_2, \ldots, P_n$ = Percentage reactives in reflectance classes 1, 2, . . ., n

P_T = Total reactives in the coal

For these indices to be useful to industry, a relationship between these parameters and a qualitative measure of coke quality is required. Many hundreds of actual coke oven tests have been run to correlate the SI/CBI values to coke strength, as measured by ASTM stability, JIS drum index, and other coke strength systems.

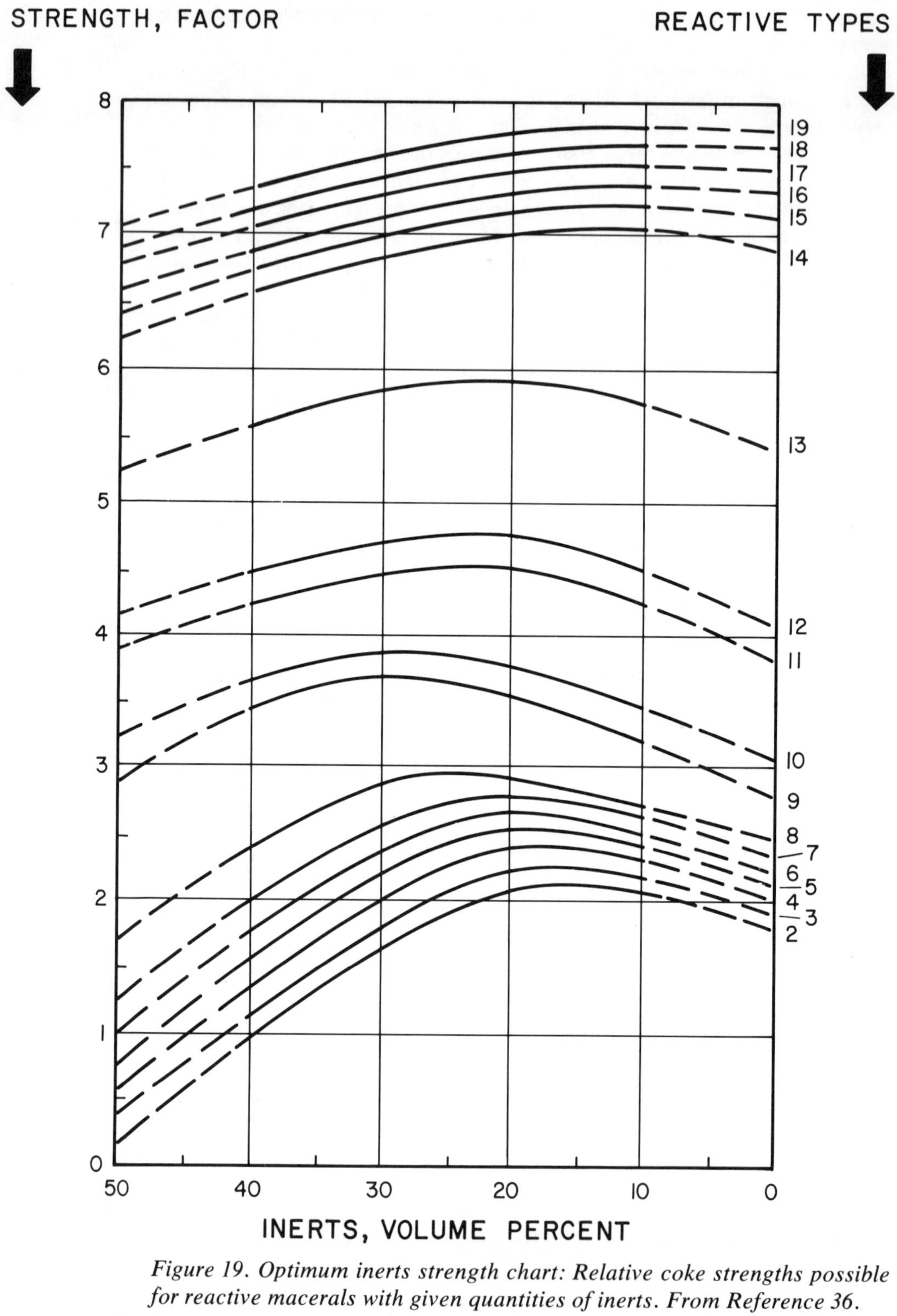

Figure 19. Optimum inerts strength chart: Relative coke strengths possible for reactive macerals with given quantities of inerts. From Reference 36.

In general, there is a correlation; however, it must be pointed out that not all coals in the world would give the same answers as shown in the Shapiro-Gray charts (see Figures 20, 21, and 24). This can possibly be attributed to errors in determining the basic maceral characteristics and other analytical errors. Some coals may have what some authorities call pseudovitrinite [42]. The standard but arbitrary designation of one-third of the semifusinites as reactive and two-thirds as inert is open to considerable question in dealing with some coals, in which the ratio has been shown to be 50 percent or more reactive. The arbitrary distribution of reactive semifusinites proportionally among the vitrinite types may be satisfactory for low-volatile coals, but does not hold true for high-volatile coals. Finally, there is the question of weathered or oxidized coals and methods for measuring effects of these processes.

In spite of these reservations, there is considerable value in evaluating the coking potential of coals by the method described, and it is being widely adopted in the industry. Its ability to show the effect of blending various grades of coal to arrive at an ideal coking coal mix has been proven.

One point of caution: the above-mentioned variations must be considered in dealing with a strange or unknown coal and proper adjustments made. The final check, of course, is the test results from the coke ovens.

Figure 20 is the basic chart developed by Shapiro and Gray and is the one commonly used in the U.S.A. and elsewhere. Isostability lines are the result of many coke oven ASTM stability tests on coals of a wide variety of reactive-inert ratios and maceral types. It will be noticed that the coke stability for any given SI is at its best where the Composition Balance Index is at 1.0. By calculating the SI and CBI, as formerly explained, the position of a particular coal can be plotted on this chart. The chart also shows the relative positions with regard to volatile range of coals of different ranks.

Figure 21 is the same chart but shows the classifications of coke properties that generally can be deduced by the position of a coal's CBI/SI location. One particularly notes that foundry coke requires high-strength coke, CBIs above 1.0 and SIs of perhaps over 4.0. Reactive chemical cokes can be relatively weak cokes of low CBI and SI.

The optimum for blast furnace metallurgical coke is as shown in the lower center of the chart, with a CBI range of 0.6 to 1.5 and an SI range of 3.5 to 4.1; an ideal blend of coals will fall into this range. Very few single coals will meet this optimum by themselves.

Figure 22 is a similar chart developed by the Japanese using isocurves of JIS, DI $\frac{30}{15}$ lines of stability instead of ASTM. It will be noticed that although the "ideal" coal is about in the same area (cross-hatched) as in the U.S.A. chart, the configuration of the isostability lines is somewhat different. This particular chart, taken from Miyazu *et al.* [12, 20], also shows the positions of various coals which are common in the Japanese coke industry.

Figure 23 is another chart by the same authors and shows the reflectance values plotted against maximum fluidities of coals. This is an extremely interesting chart, for it shows the positions of some of the most important coals used by the Japanese in the complex coking coal blends required for their modern blast furnaces. The chart also shows how fluidity affects the results.

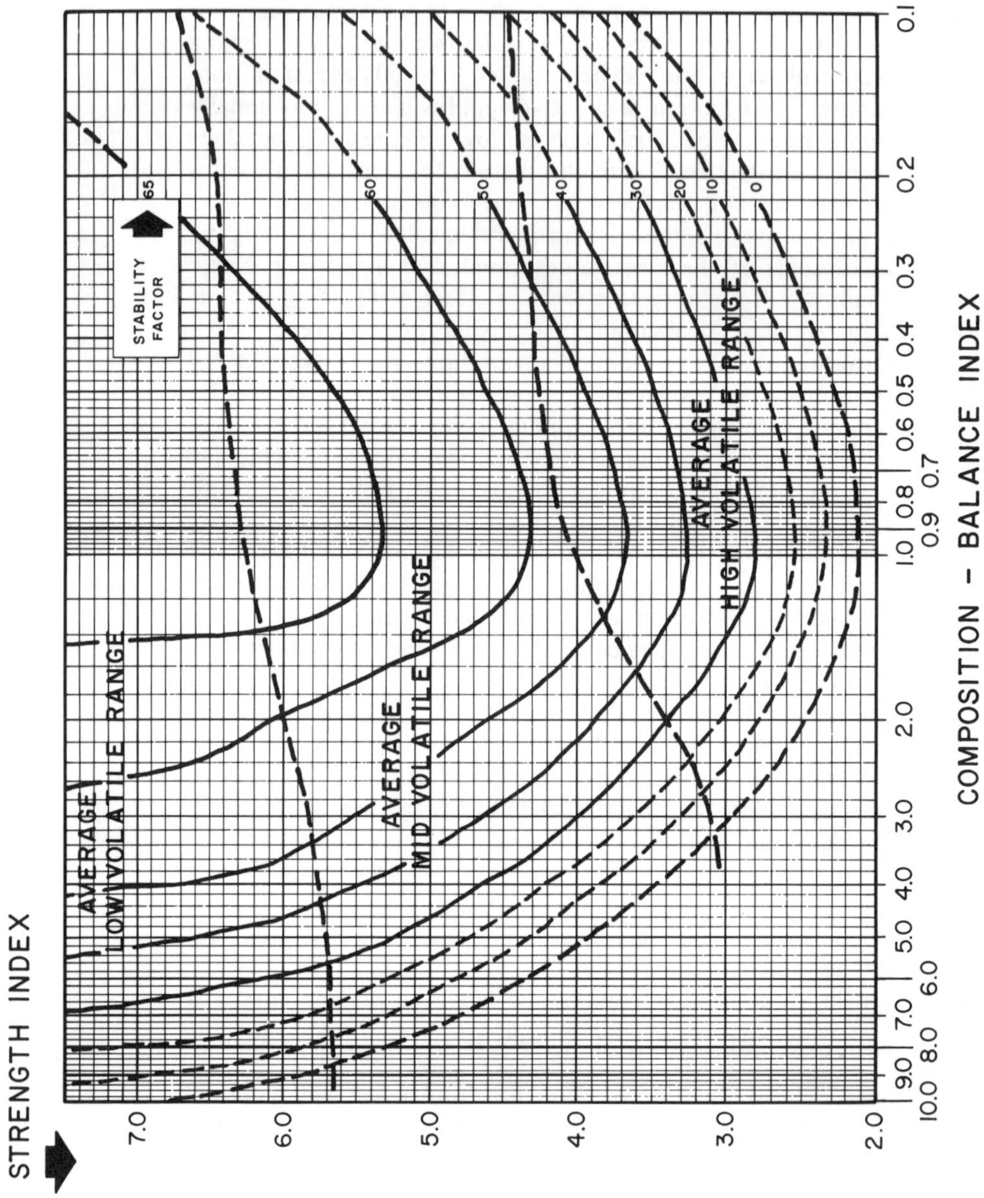

Figure 20. Relation of Strength Index (SI) and Composition Balance Index (CBI) to coke stability. From Reference 24.

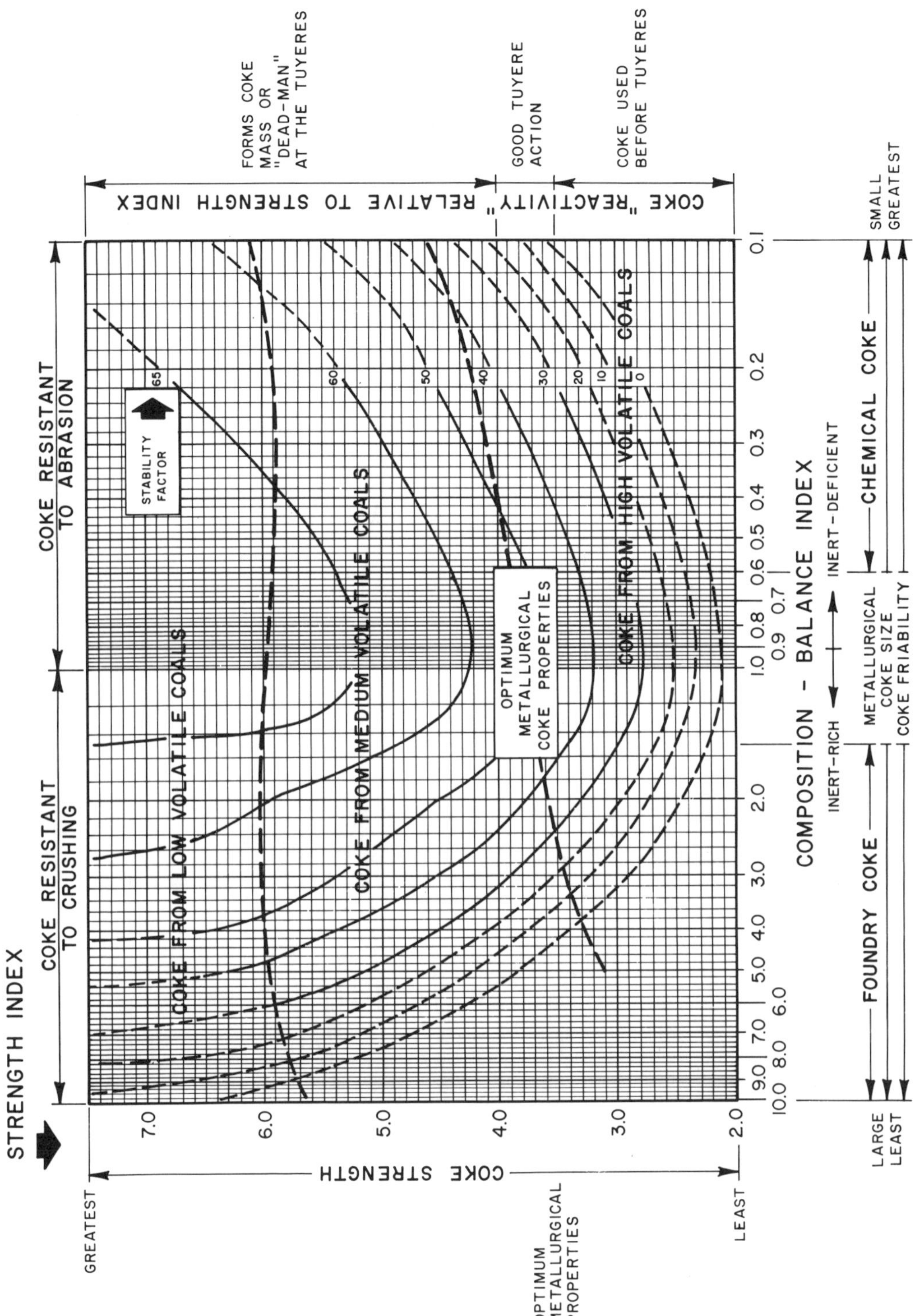

Figure 21. Characteristics of coke attainable from bituminous coals. From Reference 24.

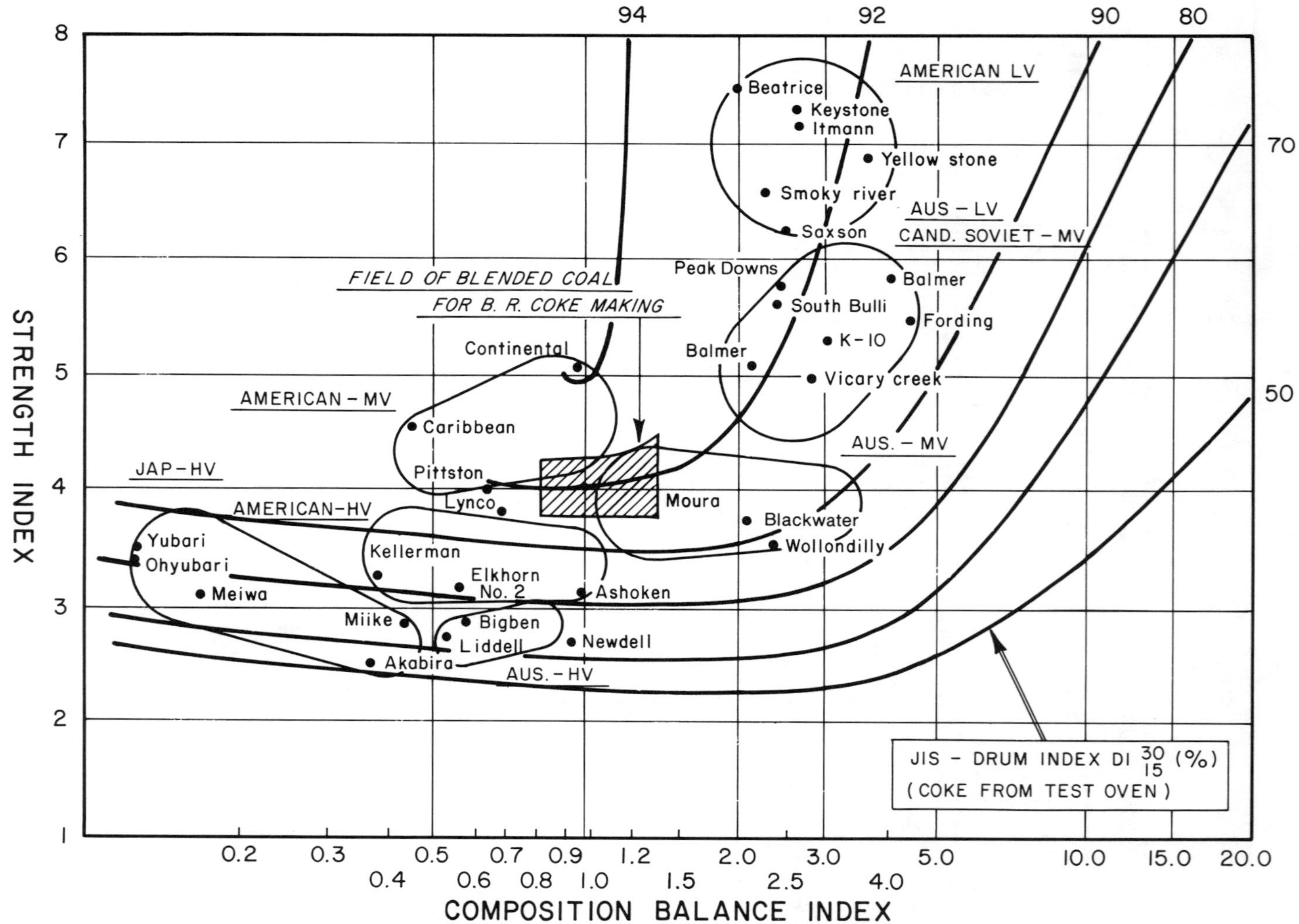

Figure 22. Relation of Composition Balance Index (CBI), Strength Index (SI), and determined DI $\frac{30}{15}$ lines. From Reference 12.

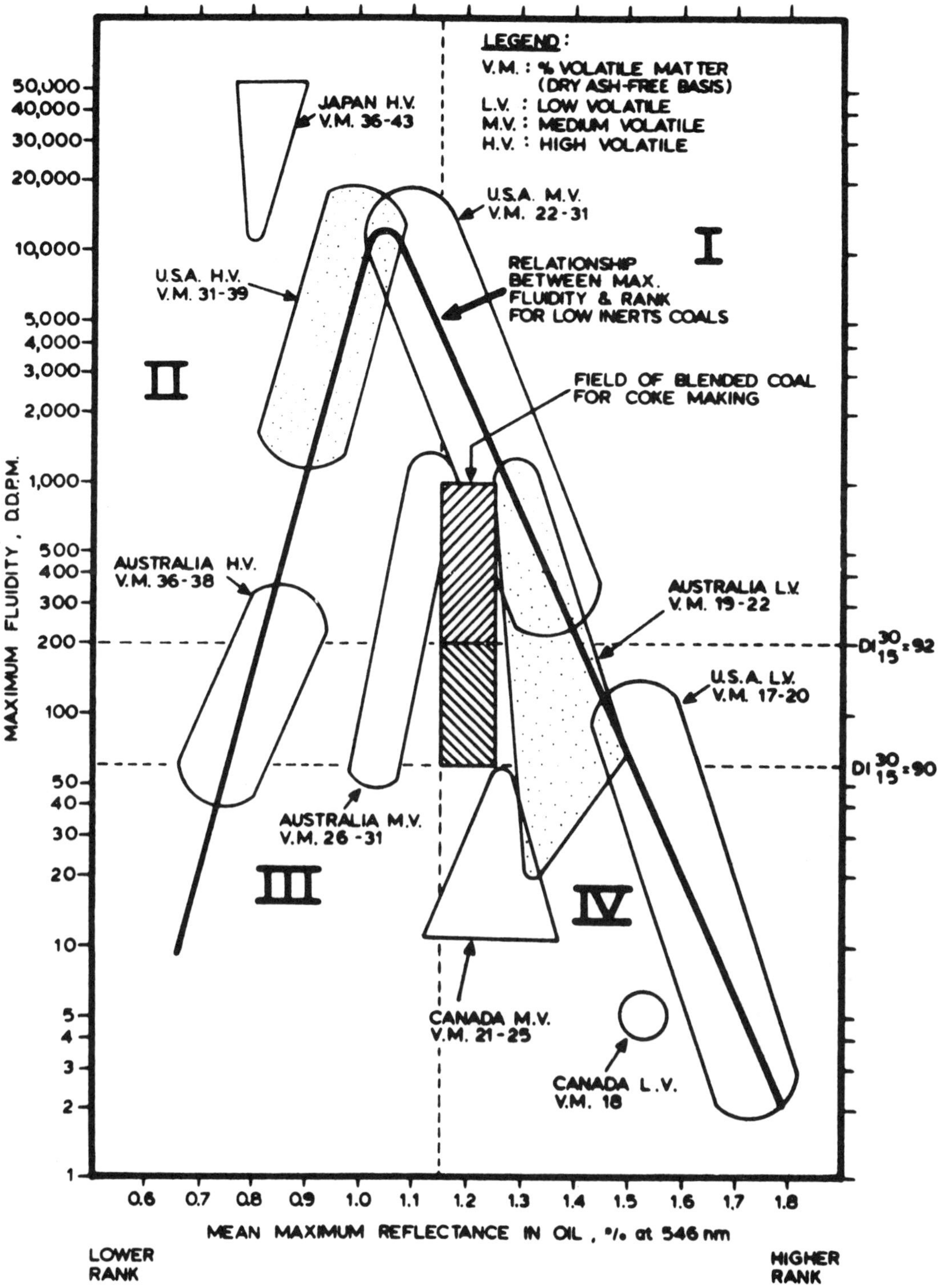

Figure 23. Relationship between maximum fluidity and mean maximum reflectance. From Reference 12.

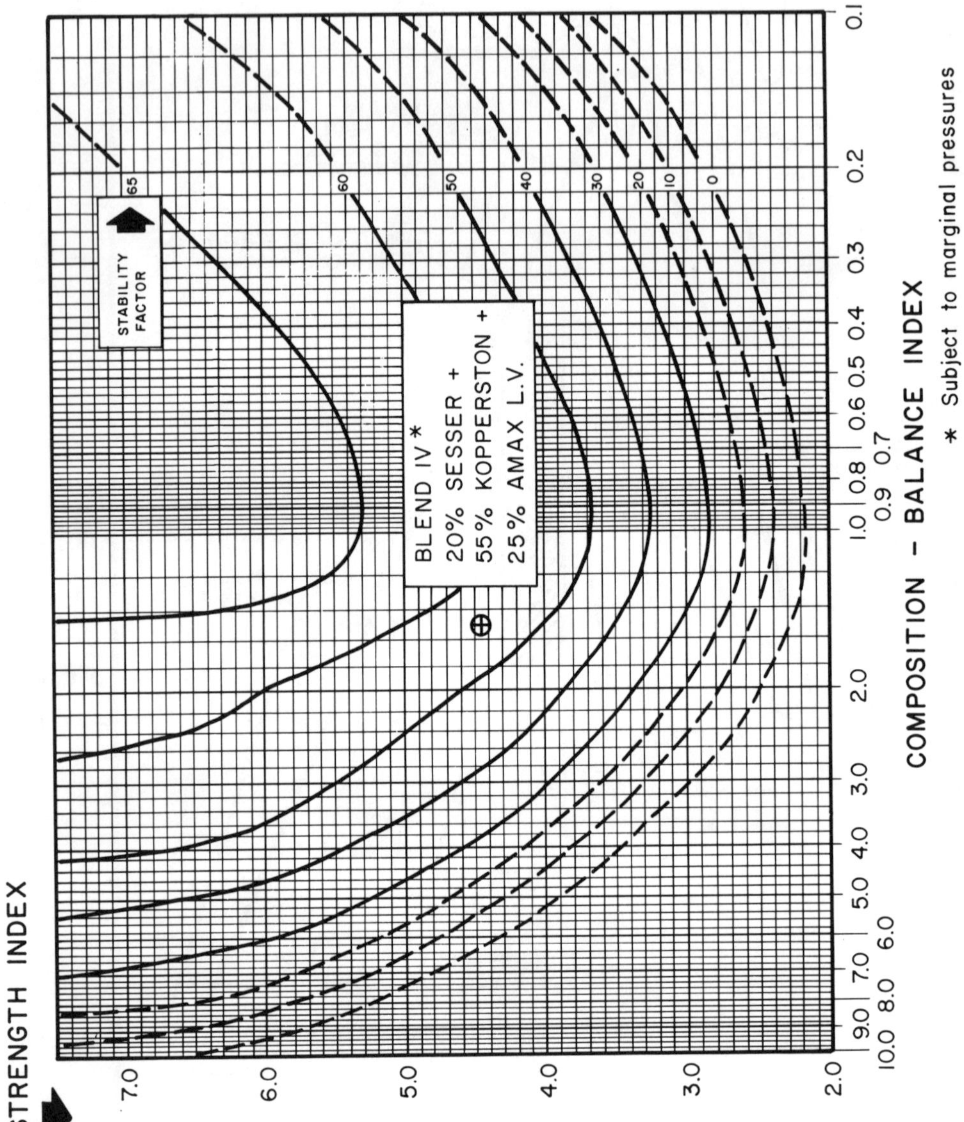

Figure 24. Blending potential of Blend VI: 20% Sesser, 55% Kopperston, and 25% Amax (Canadian) coals. From Reference 20.

Figure 24 is a chart showing the position on the Shapiro-Gray composition balance index of a blend of Seeser high-volatile coal from Illinois and Kopperston medium-volatile coal from West Virginia together with an Amax low-volatile coal from Canada. This chart indicates that by proper blending a satisfactory coke can be obtained from this mixture [31].

An interesting list of various coking coals from different parts of the world, giving pertinent chemical, fluidity, and petrographic analyses, is shown in Table 8. These are listed in a paper by Matsuoka [12]. Notice that he particularly lists the CBI and SI, together with the Japanese DI strength stability recordings.

Table 8. Properties of Typical Single Coals

	Property							
	Chemical Analysis						*Petrographic Analysis*	
Coals	*Ash*	*VM*	*T.S.*	*CSN*	*Max. Fluidity ddpm*	*Strength (DI 30/15)*	*Strength Index*	*Composition Balance Index*
American Coal								
Low Volatile Coal								
Itmann	7.5	18.5	0.67	8	48	93.4	7.11	2.84
Keystone	6.2	17.1	0.66	7½	11	93.1	7.32	2.8
Beatrice	3.9	18.3	0.67	7	10	93.1	7.47	2.0
Bird B	7.5	16.9	0.66	–½	6	92.3	7.31	1.11
Wyco	8.0	17.7	0.66	7½	65	93.0	6.92	3.07
Indian Ridge	6.3	18.6	0.69	8½	72	93.1	6.96	1.7
Special Pocahantas	7.2	20.7	0.85	8½	27	92.7	6.04	0.71
Slab Fork	7.5	17.3	0.71	5½	6	87.5	7.18	1.16
Lancashire	7.3	21.4	0.68	9	610	93.9	6.27	1.18
Middle Volatile Coal								
Pittston MV	7.3	30.0	0.67	7½	1,000	91.2	3.91	0.65
BS Blend	6.8	30.1	1.02	7	360	91.1	4.29	0.58
Jenkin Jones	6.8	23.6	0.62	8	41	92.1	6.37	1.97
Black Watch	12.1	28.1	0.76	7½	8,000	86.1	3.64	0.75
Bluestone	6.4	24.3	0.95	4½	1,000	93.7	3.69	0.78
Sewanee	8.2	29.4	0.72	6	160	93.0	2.41	0.65
High Volatile Coal								
Elkhorn No. 2	11.0	33.1	0.88	6	540	86.0	3.20	0.56
ICC HV	10.5	30.6	1.01	8	1,300	83.7	3.12	1.42
Russel Fork	7.7	32.9	0.86	6½	380	84.8	3.54	1.08
Lynco	6.2	33.2	0.80	6½	500	86.3	3.84	0.68
Masco	7.8	31.9	0.86	7½	640	84.3	2.78	0.71

Table 8. Properties of Typical Single Coals–continued

	Property							
	Chemical Analysis						Petrographic Analysis	
Coals	*Ash*	*VM*	*T.S.*	*CSN*	*Max. Fluidity ddpm*	*Strength (DI 30/15)*	*Strength Index*	*Composition Balance Index*
Australian Coal								
Coal Cliff	10.5	20.5	0.35	3½	75	93.5	5.50	2.15
South Bull	11.5	22.5	0.37	3½	470	92.0	5.50	2.42
Peak Downs	8.7	21.9	0.59	7½	250	94.8	5.72	3.05
Goonyella	7.7	25.0	0.49	6	400	94.3	4.24	1.03
Huntley	11.1	27.8	0.57	7	9,950	93.4	-	-
Blackwater	8.0	27.5	0.39	4	140	92.0	3.66	2.27
Moura	9.0	28.2	0.46	4½	530	91.2	3.89	1.45
Wollondilly	10.9	27.5	0.39	3½	135	91.9	3.40	2.50
Newdell	9.2	38.3	0.49	4½	345	80.0	2.75	0.95
Liddell	8.7	38.5	0.64	4	230	73.3	2.76	0.56
Big Ben	10.0	36.3	0.69	4	100	77.8	2.85	0.58
Abersea	9.8	37.8	0.74	2½	19	71.0	2.95	1.08
Lithgow	14.0	31.4	0.80	1	7		2.00	2.93
Canadian Coal								
Smoky River	7.5	19.2	0.44	8	15	93.2	6.54	2.41
Balmer	11.4	20.8	0.27	5	10	89.1	5.0	2.1
Fording	8.5	21.4	0.37	6	17	92.7	5.42	4.25
Luscar	8.8	21.1	0.25	5	23	91.2	5.6	2
Vicary Creek	10.1	22.7	0.56	4	29	85.0	4.97	2.92
Soviet Coal								
Kuznetsk KJ-14	7.7	24.9	0.36	3	40	92.5	-	-
Kuznetsk K-10	7.7	21.3	0.38	2	4	88.8	5.38	3.04
Kuznetsk G-6	7.2	39.0	0.64	2½	250	49.6	3.02	0.43
Kusnetsk OS	8.4	14.6	0.40	1½	-	38.4	6.26	9.38
Other Foreign Coal								
1 st. Maja	6.7	28.8	0.82	7	680	91.7	4.20	1.30
Marcel	6.6	34.5	0.76	4	10	90.0	4.09	0.58
Wittbank	7.7	32.2	0.64	1	6	86.3	3.73	4.46
Domestic Coal								
Yubari	6.7	30.4	0.36	8	27,500	78.6	3.65	0.11
Helwa	6.5	40.0	0.30	8	27,500	83.7	3.10	0.14
Ohyubari	6.9	39.3	0.34	9	27,500	79.0	3.50	0.11

Table 8. Properties of Typical Single Coals–continued

	Property							
	Chemical Analysis						Petrographic Analysis	
Coals	*Ash*	*VM*	*T.S.*	*CSN*	*Max. Fluidity ddpm*	*Strength (DI 30/15)*	*Strength Index*	*Composition Balance Index*
Domestic Coal—continued								
Ashibetsu	6.7	41.0	0.57	5	120	62.0	2.57	0.22
Akahira	6.6	40.1	0.77	4½	500	58.7	2.44	0.36
Takashima	6.9	40.8	0.71	7½	27,500	72.6	2.78	0.13
Miike	7.6	41.7	1.70	9	27,500	72.6	2.83	0.43

Source: Reference 12.

6 Contract evaluations

Different users have different contract requirements for purchasing coking coals. Quite often, a great deal depends upon market conditions—the law of supply and demand. Existing price schedules in the current world market vary considerably, for the same reason. Generally, low-volatile coals demand higher prices than high-volatile coals, and medium-volatile coals are somewhere in between. However, there is currently an increasing demand for high-fluid coals, especially to counteract the use of coal briquets and high percentages of non-reactive coals.

Transportation costs from point of production to point of use are of vital concern; also affecting contract terms (and price) are the quantity of coal to be purchased and whether it is a short-term or long-term contract. The user requires assurance as to the ability of the producer to operate satisfactorily, to ensure a steady supply of coal of uniform consistency. The buyer will (or should) require adequate proof that the producer has sufficient uncommitted reserves of satisfactory quality to service the contract.

Other contract considerations to be determined and defined include items such as:

- Are the prices quoted on the basis of f.o.b. mine, rail, barge, or ship-loading point, or are they c.i.f. destination?
- What are the escalation factors to be agreed upon?
- What type of "force majeure" situations may arise?
- Are unit trains involved?
- If ocean transport is involved, what are the maximum ship capacities that can be handled?
- What storage facilities are available at loading and unloading points and what are their capacities?

And then there are the payment conditions, including penalty and bonus clauses; methods and point of sampling; and whose analyses are to be taken as official, etc. All of these have to be spelled out in detail.

Definite limits are placed on such quality characteristics as:

Moisture	Sulfur
Ash	Phosphorus
Volatile Matter	Free-Swelling Index (Crucible-Swelling Index)

Gieseler fluidity value is an increasing consideration, as are elements of a deleterious nature.

Ash is generally quoted on a ± 0.5 percent tolerance, with a price penalty for exceeding the "base" contract percentage. (A "bonus" is seldom given for being under the base percentage.)

Typically, it is required that the sulfur should not be more than 1.0 percent, with a 0.05 percent tolerance. There can be exceptions to this formula in some cases.

Guaranteed analyses may be on an "as-received," "air-dryed," or "dry" basis. Quite often the ash, volatile matter, and sulfur contents are on a dry basis, but shipment tonnage is on a maximum allowable moisture basis (generally 6 to 8 percent). Spelling this out is important for administering a contract.

Where the fluidity of a coal is of vital importance, such as in establishing a base price for its value, some contracts may include a minimum Gieseler fluidity requirement in the specifications. Similarly, Audibert-Arnu dilatometer (Ruhr or IRSID), as well as the Roga index, may be inserted in European contracts.

Generally, the price of a coking coal is calculated by adjusting the price of a reference coal for differences in ash, sulfur, volatile matter, and other parameters [43]. Price adjustments for ash differences may vary from 2 to 4 percent of the price per 1-percent difference in ash. Japanese mills currently use a flat $2.00 (U.S.) per tonne adjustment for 1-percent ash. (The word *tonne* refers to metric tons, rather than the long or short tons used in U.S. practice.)

For sulfur, factors commonly used for price adjustments are 5 to 6 percent of the price per 1-percent difference in sulfur. Japanese mills allow $10.00 (U.S.) per tonne per 1-percent sulfur. The European Coal Commission uses 5 percent multiplied by the sulfur recovery in coke.

Price adjustments for volatile matter depend upon values recovered from by-products and range from 0.4 percent to 0.9 percent of the price per 1-percent volatile matter. Japanese mills allow $0.30 to $0.40 (U.S.) per tonne per 1 percent (dry basis)

Price differences related to moisture vary, since transportation costs must be accounted for and the accepted basic moisture level may differ.

On petrographic analysis, the Japanese may use two methods for price evaluations: one is based upon the relationship of the Strength Index (SI) to the Composition Balance Index (CBI), and the second is based on the relationship between maximum fluidity and mean reflectance.

For the SI/CBI relation to the JIS drum index (DI) $\frac{150}{15}$, the Japanese currently place a value of about $1.25 (U.S.) per tonne on each 0.1-percent variation in DI $\frac{150}{15}$.

Using the second method, fluidity/mean reflectance (Ro), a basic value, may be expressed as:

$$C_1 \times R_o (1 - \text{inert}) + C_2 \times \text{Inert} + C_3 \times \log MF$$

where:

C_1, C_2, C_3 = Dollar constants (C_2 is the price of a standard non-coking coal.)
R_o = Mean reflectance
MF = Maximum fluidity

This basic value is then adjusted for differences in chemical properties to arrive at a final value.

These are but some of the merit rating formulas. Such evaluations can become quite complex. Depending upon how close price bargaining is and, in the final analysis, the actual needs of the buyer for a particular coal, many other systems can be employed.

References

1. U.S. Bureau of Mines, Bulletin 638, *Methods of Analyzing and Testing Coal and Coke,* 1967.

2. American Society for Testing and Materials (ASTM), *Standards for Coal and Coke,* Committee D-5, 1975-1976 (and subsequent supplements). Documents are available from the ASTM, 1916 Race Street, Philadelphia, Pennsylvania 19103.

3. International Organization for Standardization (ISO). A worldwide federation of national standard institutes, supported as an agency of the United Nations Economic Commission for Europe. Prior to 1972, the results were published as recommendations. Since that date, various tests have become "standard," a continuing process.) Documents generally available from the American National Standards Institute, 1430 Broadway, New York, New York.

4. British Standards Institution (28 Victoria Street, London, S.W. 1, England), British Standard 735, *Sampling and Analysis of Coal and Coke.*

5. Japanese Industrial Standards (JIS), *Standards for Coal and Coke.* Various bulletins on all phases, published by the Japanese Standards Association (in English), 89 Hitotsugi-Cho, Akasaka, Minato-Ku, Tokyo, Japan.

6. United Nations Economic Commission for Europe, *International Classification of Hard Coals by Type,* E/ECE/247/Coal/110 (1956). See Reference 3.

7. U.S. Bureau of Mines, Report of Investigations 5435, *The International Systems of Hard Coal Classification and Their Application to American Coals,* 1958.

8. U.S. Bureau of Mines, Report of Investigations 5695, *The International System for Classifying Brown Coals and Lignites and Its Application to American Coals,* 1960.

9. Miyazu, T., T. Fukuyama, and T. Koizumi, *Effect of Coking Properties on Blast Furnace Operation,* N.K.K. Nippon Kokau, Tokyo, Japan, Technical Report, December 1970.

10. Fisher, A. D., Vice President, The Steel Company of Canada, Ltd., *Metallurgical Coals for the Future,* Twenty-Third Canadian Conference on Coal, Ottawa, Ontario, Canada, September 1971.

11. Clendenin, J. D., U.S. Steel Corporation, *The Third Carbonization Science Lecture of the Coke Oven Managers Association of Great Britain,* London, England, November 3, 1971.

12. Matsuoka, H., Nippon Steel Corporation, *Requirements for Coals in Japanese Coking Blends,* in, Cook, A.C. (ed.), *Australian Black Coal—its occurrence, mining, preparation and use,* Australasian Institute of Mining and Metallurgy, Illawara Branch:10-18.

13. Montgomery, C. R., and L. J. Verno, Clinchfield Coal Company, Dante, Virginia, *Observations in Evaluating Coals for Carbonization,* AIME (SME) Preprint 62F82, 1962.

14. Rees, O. W., Illinois State Geological Survey, *Chemistry, Uses and Limitations of Coal Analyses,* Report of Investigations 220, Illinois State Geological Survey, Urbana, Illinois, 1966.

15. Le Page, A. J., B. W. Proudfoot, and J. B. Sedgman, Australian Coal Industry Research Laboratories, Ltd. (ACIRL), *Use of Laboratory Data for Preparation Plant Assessment,* P.R. 74-3, P.R. 76-7, P.R. 77-7, Chatswood, N.S.W., Australia, 1974, 1976, and 1977.

16. Leonard, J. W., University of West Virginia, Morgantown, West Virginia, *Advanced Work on the Determination of Coke Stability from Coal, Hardgrove Grindability Index, Bulk Density and Volatile Matter,* AIME, Preprint 65F1, February 1965.

17. Martindill, G. H., and M. J. Kovalik, U.S. Bureau of Mines, Report of Investigations 5699, *The Role of Fluidity in Carbonization,* 1960.

18. U.S. Bureau of Mines, Information Circular 8010, *Comparative Evaluation of Coking Coal Properties of Four Coals,* 1961.

19. Wu, W.R.K., U.S. Bureau of Mines, Report of Investigations 7394, *Repeatability of Plastometer and Dilatometer Tests of Coal,* 1970.

20. Miyazu, T., Y. Okuyama, H. Sugimura, and M. Kumagai, Journal of the Fuel Society of Japan, *Prediction of the Coking Property of Coal by Microscopic Analysis,* Vol. 1., 1970.

21. Kojima, K., Iron and Steel International, *Automatic System for Evaluating Coking Coals,* 1976.

22. Deutscher Normenausschuss, DIN51-739, *Testing of Solid Fuels: Determination of Coking Capacity of Bituminous Coals by the Dilatometer,* 1976.

23. Gray, R. J., A. H. Rhoades, and D. T. King, SME of AIME, Preprint 74-F-61, *Detection of Oxidized Coal and the Effect of Oxidation on the Technological Properties,* 1974.

24. Shapiro, N., R. J. Gray, and G. R. Eusner, Iron and Steel Division, AIME Proceedings, *Recent Developments in Coal Petrography,* Vol. 20, 1961.

25. Schulze, W. G., and G. C. Soth, AIME, Blast Furnace, Coke Oven and Raw Materials Conference, Chicago, Illinois, *Use of Coal Plasticity in Coke Plant Blending,* April 1960.

26. Shapiro, N., and R. J. Gray, U.S. Steel Corporation, Illinois Mining Institute Proceedings, *Petrographic Constituents of Coal,* 1960.

27. Edwards, G. E., and A. C. Cook, Proceedings, Australasian Institute of Mining and Metallurgy, *Blend Design for Metallurgical Coke Production,* No. 244, December 1972.

28 Harrison, J. A., Illinois Mining Institute, Proceedings, *Coal Petrography Applied to Coking Problems,* 1961.

29. Ammosov, I. L., I. V. Erneu, S. I. Sukhenko, and L. S. Oshurkova, Koks i Khimiyn, U.S.S.R., No. 12, *Calculation of Coking Charges on the Basis of Petrographic Characteristics of Coals,* 1957.

30. Abramski, C., M. Th. Mackowsky, W. Mantel, and E. Stach, Verlag Gluckhauf, Essen, Germany, *Atlas fur Angewadte Steinkohlenpetrographie,* 1951.

31. Harrison, J. A., H. W. Jackman, and J. A. Simon, Illinois State Geological Survey, Circular 366, *Predicting Coke Stability from Petrographic Analyses of Illinois Coal,* 1964.

32. Berry, W. F., A. R. Cameron, and B. N. Nandi, Centennial Reviews, CIM, Ottawa, Canada, *The Development of Coal Petrology in North America,* 1967.

33. Schoenberger, R. W., and B. R. Kuchta, U.S. Steel Corporation, AIME Iron-Making Conference, Toronto, Ontario, Canada, *Design and Application of an Adjustable-Width, Movable-Wall-Type Oven,* 1961.

34. Gayle, J. B., and W. M. Eddy, U.S. Bureau of Mines, Report of Investigations 5744, *Carbonizing Tests with the Tuscaloosa Oven,* 1961.

35. Price, J. G., R. W. Schoenberger, and B. Perlic, AIME, *Use of Thirty-Pound Oven for Rapidly Assaying the Coking Strengths of Coals,* 1958.

36. Moses, R. G., Bituminous Coal Research Inc. (350 Hochberg Road, Monroeville, Pennsylvania), *The Use of Petrographic Analysis for the Prediction of Coke Stability,* 1976.

37. Cole, D. L., and W. F. Berry, Bituminous Coal Research Inc., (350 Hochberg Road, Monroeville, Pennsylvania), *Preparation and Polishing of Coal and Coke for Petrographic Analysis,* 1965.

38. International Committee for Coal Petrology, published by Cerchar, Paris, France, *Glossary of Coal Petrological Terms,* 1957.

39. Mackowsky, M. Th., Proceedings, International Committee for Coal Petrology 2, *Neu Wege Zur Petrographischen Kennzeichnung von Kokskohlen,* 1955.

40. Parks, B. C., Chemistry of Coal Utilization, Supplemental Volume, H. H. Lowry (John Wiley & Sons, Inc.), New York, New York, *Origin, Petrography and Classification of Coals,* 1963.

41. International Committee for Coal Petrology, *International Handbook of Coal Petrography,* 2nd Edition, published by Cerchar, Paris, France, 1963, or later editions if published.

42. Benedict, L. G., R. R. Thompson, J. J. Shigo, III, and R. P. Aikman, *Fuel,* A Journal of Fuel Science, Butterworth's Scientific Publications, London, England, *Pseudovitrinite in Appalachian Coking Coals,* 1968.

43. Bateman, K. W., and W. Coulter, The Australian I.M.M. Symposium, held in Newcastle, N.S.W., Australia, *Commercial Evaluation of Coking Coal from Borehole Data,* Oct.-Nov. 1977.

44. Mutschler, P. H., U.S. Department of Interior, Information Circular 8677, *Impact of Changing Technology on the Demand for Metallurgical Coal and Coke Produced in the United States to 1985,* 1975.

Appendices

A Hard coal—Audibert-Arnu dilatometer test

B Ruhr dilatometer

C Coal—determination of caking power—Gray-King coke test

D Hard coal—determination of caking power—Roga test

E Second draft proposal for the Micum test
by ISO Technical Committee 27, Solid Mineral Fuels

Supplement: Chairman's report on
the work of ISO Technical Committee 27

F Japanese Industrial Standard drum and tumbler test

G Sole-heated oven dilatometer for measuring the expansion and contraction
behavior (volume change) of coals

H Adjustable-width oven illustrations

Appendix A

Hard Coal—Audibert-Arnu Dilatometer Test
ISO Standard 349-1975 (E)

This material is reproduced with permission from the International Organization for Standardization, Standard 349-1975(E) "Hard Coal—Audibert-Arnu Dilatometer Test," copyrighted by the American National Standards Institute, 1430 Broadway, New York, New York 10018.

Hard coal – Audibert-Arnu dilatometer test

0 INTRODUCTION

The Audibert-Arnu test is one of the parameters adopted for the International Classification of Hard Coals by Type of the United Nations Economic Commission for Europe. The object of the test is to determine the coking properties of hard coal or hard coal blends on the laboratory scale.

In principle, the test is not designed, nor can it be used, to indicate the pressures exerted by hard coals on the walls of industrial carbonization ovens.

1 SCOPE AND FIELD OF APPLICATION

This International Standard specifies a method for determining the swelling properties of hard coal when heated under standard conditions in a dilatometer.

2 PRINCIPLE AND TERMINOLOGY

A pencil made of powered coal formed under pressure is inserted into a precisely calibrated narrow tube and topped by a calibrated steel rod (piston) which slides in the bore of the tube.

The whole is heated at a constant and definite rate.

By making regular readings of the displacement of the piston as a function of the temperature and expressing the displacements observed as percentages of the original length of the pencil, a curve of the type shown in figure 1 can be plotted.

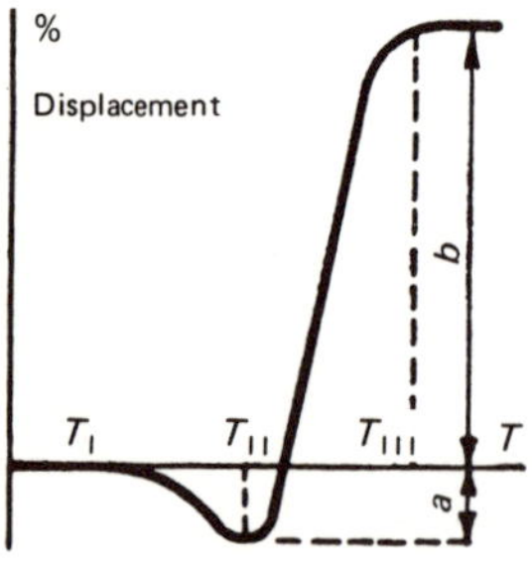

FIGURE 1

The following items are characteristic (see note below) :

T_I temperature at which the piston has moved down 0,5 mm[1] : *softening temperature.*

T_{II} temperature at which the piston reaches its lowest point : *temperature of maximum contraction.*

T_{III} temperature at which the piston reaches its highest point : *temperature of maximum dilatation.*

a maximum contraction of length of pencil, per cent.

b maximum dilatation of length of pencil, per cent.

If, after contraction has taken place, the piston does not return to its original level, the dilatation, equal, in absolute value, to the difference between the final level of the piston and the original zero level, is reported as negative.

NOTE – The principal factors capable of distorting the results of this empirical test are the following :

a) Deterioration of the coal, consequent on bad storage or faulty drying;

b) Deviation from the tolerances of

1) the internal dimensions of the dilatometer tube,

2) the clearance between tube and piston,

3) the mass of the piston,

4) the dimensions of the mould;

c) Deviation from the specified mean rate and regularity of heating;

d) Deviation from the specification for the preparation of the sample in respect of maximum particle size, or for the pencil in respect of its length after tamping

3 APPARATUS

3.1 Apparatus for preparing the coal pencil

3.1.1 Mould, polished internally, with accessories; see figures 2 and 2A.

3.1.2 Gauge, see figure 2.

3.1.3 Ram, of which figure 3 shows an example.

3.1.4 Press, of which figure 4 shows an example.

1) Or one division, if the scale is calibrated in percentage of the standard length of pencil.

3.2 Dilatometer and accessories

3.2.1 Dilatometer tubes and pistons, see figure 5.

3.3 Other apparatus

3.3.1 Electric furnace, of which figure 6 shows an example, consisting of a block of metal resistant to oxidation and of sufficiently high melting point. A suggested material is aluminium bronze. The metal block is pierced by at least two holes of 15 mm diameter by 350 mm deep to take the dilatometer tubes and one hole 320 mm deep to take a temperature-measuring device. The block is heated by a metallic winding, suitably insulated. Control gear permits the use of any selected rate of heating up to 5 °C per minute over a temperature range of 300 to 550 °C.

The furnace shall be constructed so that the temperature conditions are uniform in the dilatometer tubes placed in their normal position in each hole.

To verify this, heat the furnace at a rate of 5 °C per minute. When the temperature reaches about 450 °C, make an exploration of the lower 180 mm of a dilatometer tube placed in the furnace, by comparing the readings on the normal temperature-measuring device and a thermometric probe placed in the tube. The difference between the probe temperatures and the mean temperature shown by the normal temperature-measuring device shall be less than

± 2 °C in the lower 120 mm,
± 5 °C from 120 to 180 mm

This verification is not to be confused with the later calibration of the temperature-measuring device; it is intended to measure only the temperature variation along the tube.

The furnace shall be equipped with an adjustable scale for each hole. The scale shall be preferably engraved on a mirror in front of which the indicator pointer of the piston moves. It shall be at least 180 mm in length and calibrated in millimetres or in percentages of the standard length of the pencil (60 ± 0,25 mm, see 5.1).

If desired, the apparatus can be equipped with an automatic heating regulator and a device for the automatic registration of the curve.

3.3.2 Temperature-measuring device, consisting of a mercury thermometer, a thermocouple or a resistance thermometer, accurate to within 1,0 % of the temperature in degrees Celsius and capable of being read, if necessary by estimation, to 1 °C.

3.3.3 Cleaning implements, consisting of the following :

3.3.3.1 Auger, diameter approximately 7,8 mm;

3.3.3.2 Reamer, consisting of a steel bar of semi-circular section of diameter 7,95 mm;

3.3.3.3 Brass wire brush, the diameter of which shall slightly exceed 8 mm.

The total length of each of the cleaning implements shall be 400 mm.

3.4 Calibration of apparatus

Calibrate the apparatus by comparing the temperature in a dilatometer tube in each hole with the temperature indicated by the temperature-measuring device in its normal position. Carry out the calibration at the desired rate of heating by using a thermocouple with wires of diameter approximately 0,6 mm, the thermojunction touching the wall of the tube 30 mm above the bottom. Correct the temperatures read during the test by the differences found during this calibration.

3.5 Inspection

3.5.1 *Dilatometer*

In order to inspect the wear of the tube and piston after a hundred determinations have been carried out in one tube, compare the results of the next four determinations using that tube with those obtained in a new tube. This comparison will thus be made successively on four coals.

Divide the difference in percentage dilatation between the two tubes by the "relative length" of the dilated pencil obtained with the new tube; the "relative length" is here expressed as the ratio of the length of the dilated pencil to its original length.

Average the figures so obtained from the four coals. If the average is greater than 3,5, irrespective of sign, discard the old tube (see annex). If the tube is still satisfactory, repeat this comparison after every subsequent 25 tests.

3.5.2 *Mould*

Check the wear of the mould periodically with the gauge, which can also be used to check new moulds.

If, when the gauge is inserted in the larger orifice of the mould,

1) two lines can be seen on the gauge, the mould is too small and shall be reamed out;

2) one line can be seen, the mould is satisfactory;

3) no line can be seen, the mould is worn and shall be replaced.

4 PREPARATION OF SAMPLE

As certain types of hard coal are very susceptible to oxidation, it is necessary to minimize the contact with air after reducing the gross sample.

As a special precaution, therefore, store the test sample after reduction in an oxygen-free nitrogen atmosphere or in freshly boiled water. In the latter case, make a paste of the coal with water and put the paste into a flask which is then filled to the top with freshly boiled water.

Care shall be taken to ensure that the test sample taken is truly representative.

A suitable relationship between the mass of the test sample and the maximum particle size is shown in the following table :

Maximum particle size	Minimum mass of test sample
mm	g
5	1 000
4	500
3	250
2	100
1,5	50

If the coal has been stored under water, filter on a suction filter. Expose the coal on the filter paper at a temperature not exceeding 40 °C until the coal appears to be dry, but in any case for not more than 2 h.

Reduce the maximum particle size to 1,5 mm. Mix and take a part sample of 50 to 100 g. Crush to pass a 0,16 mm mesh sieve. Both crushing operations shall be controlled so as to produce the minimum of fines (see note). Mix again and carry out the determination on an average sample of about 10 g. Moisten this sample with 1 ml of water and mix rapidly. Too intensive mixing is liable to cause difficulty when the pencil is removed from the mould. For the same reason, it is essential that the preparation of the pencil shall be carried out without interruption.

NOTE – Too fine grinding of the coal affects the result of the determination The sample shall be crushed to obtain the following size analysis :

through 0,2 mm	100 %
through 0,1 mm	85 to 70 %
through 0,06 mm	70 to 55 %

5 PROCEDURE

5.1 Preparation of the coal pencil

Place the mould on its support with the larger orifice upwards and set the funnel on the mould. Place the coal in the funnel and lightly tamp into the mould without moving the funnel, by means of a tamping pin. Place the mould assembly under the ram in order to tamp the sample by dropping the plunger three or four times until the mass of coal ceases to yield. Repeat this three or four times until the mould is filled.

In order to remove the coal pencil from the mould, remove the support and the funnel. Place the ejector guide at the end of the mould corresponding to the smaller diameter of the pencil. Place the guide tube at the other end of the mould and the receptacle in the guide tube. Then insert the ejector piston in the guide and push the coal pencil onto the receptacle by means of the press (see note).

Then adjust the length of the pencil to 60 ± 0,25 mm by cutting away as much as necessary of the thick end with a fine blade.

NOTE – Particularly when dealing with coals which are difficult to remove from the mould, it is recommended that the ejector piston be removed from time to time and cleaned, the inner surface of the mould being cleaned at the same time.

5.2 Determination of dilatation

Heat the coal at a rate of 3 °C per minute.

Carefully insert the pencil, thick end first, into the dilatometer tube and push it very gently into position with the piston.

Place the tube and contents in one of the holes of the metal block, when the temperature of the furnace is 330 °C. Place in any hole in the block which is not being used an empty tube complete with its piston. Where, in exceptional cases, T_I is less than 350 °C, charging of the furnace shall take place when the temperature is 20 °C below T_I.

After the charged dilatometer tube has been inserted in the furnace, wait for the indicator pointer of the piston to reach a position of equilibrium before adjusting the zero of the scale. This position should be reached after about 5 min.

Immediately the dilatometer is placed in the furnace, the temperature begins to drop, and heating shall be regulated so as to regain the temperature of 330 °C at the end of 7 to 10 min.

After 340 °C has been reached, the rise in temperature shall be very steady, minute by minute, and equal to the rate stated, with a tolerance of ± 3 % of the specified temperature rise in a 5 min period (see note).

During each 5 min period, the operator shall adjust the heating rate to correct any deviation observed in the preceding period, in order to avoid the accumulation of errors.

If the curve is not automatically recorded, note the time, the position of the piston and the temperature at intervals of not greater than 5 °C. In the region of the critical points, sufficient points shall be plotted to determine the exact shape of the curve.

Continue heating for 5 min after the maximum dilatation is attained. Then stop heating and immediately remove the piston, in order to prevent its getting jammed.

Carry out the duplicate determination in a separate run.

NOTE – The tolerance stated (± 3 % of the specified temperature rise in a 5 min period) may not be attainable if the temperature-measuring device in use is such as to require a change of 1 °C to be estimated rather than read directly. In such cases, a tolerance of ± 1 °C per 5 min is recommended.

5.3 Cleaning of the tube and piston

It is essential that the test be carried out with the piston and the dilatometer tube scrupulously clean. The following method of cleaning is recommended :

5.3.1 *Tube*

Crush the semi-coke and remove as much of it as possible with the auger. Then fill the tube with crude benzene or other appropriate solvent and allow to soak for several hours. Complete the cleaning with the reamer, ensuring that no solid remains at the bottom or on the wall. Immediately before the test, clean finally with the brass wire brush.

5.3.2 *Piston*

Clean the piston, including the base, with very fine emery paper, taking care not to round the edges, and check that the piston slides freely in the tube.

6 EXPRESSION OF RESULTS

Calculate the observed changes in length as percentages of the initial length of the pencil.

7 PRECISION OF DETERMINATION

	Maximum acceptable difference between results	
	Same laboratory (Repeatability)	Different laboratories (Reproducibility)
Dilatation	$7\left(1 + \frac{b}{100}\right)$ where b is the maximum dilatation per cent	(see 7.2)

7.1 Repeatability

The results of duplicate determinations, carried out at different times in the same laboratory by the same operator with the same apparatus on the same analysis sample, should not differ by more than the above value.

7.2 Reproducibility

No value for reproducibility can be quoted for determinations carried out in different laboratories since insufficient evidence is available on which to base such a value.

8 TEST REPORT

The test report shall include the following particulars :

a) whether the sample submitted to the test fulfils the conditions specified in clause 5;

b) the rate of heating employed;

c) a curve of the percentage changes in length as a function of temperature on a standard relative scale in which 10 °C on the horizontal scale is equal to 5 % on the vertical scale;

d) the corrected temperatures T_{I}, T_{II} and T_{III} rounded off to the nearest 5 °C;

e) the percentage contraction (a) rounded to the nearest whole number;

f) the percentage dilatation (b) rounded :

to the nearest whole number for negative dilatation;

to the nearest 5 % for positive dilatation up to 100 %;

to the nearest 10 % for positive dilatation over 100 %.

ANNEX

INSPECTION OF DILATOMETER TUBES

After 100 determinations have been carried out in one tube, the next four determinations made shall be duplicated in a new tube. The results shall then be examined as follows :

If b_o is the dilatation with the old tube, and

b_n is the dilatation with the new tube,

calculate the ratio : $$x = \frac{b_o - b_n}{1 + \frac{b_n}{100}}$$

Then, if the average value of x for the four coals, taking into account the sign, is greater than 3,5 irrespective of sign, the old tube shall be rejected.

Example 1

Coal	b_o	b_n	$b_o - b_n$	x
A	100	113	− 13	− 6,1
B	13	17	− 4	− 3,4
C	61	59	2	1,3
D	45	55	− 10	− 6,4

The average value of x = − 3,6

The old tube shall therefore be rejected.

Example 2

Coal	b_o	b_n	$b_o - b_n$	x
E	54	56	− 2	− 1,3
F	81	80	1	0,6
G	109	117	− 8	− 3,7
H	40	44	− 4	− 2,8

The average value of x = − 1,8

The old tube may therefore be used for a further 25 determinations, after which it shall be re-inspected.

Dimensions in millimetres

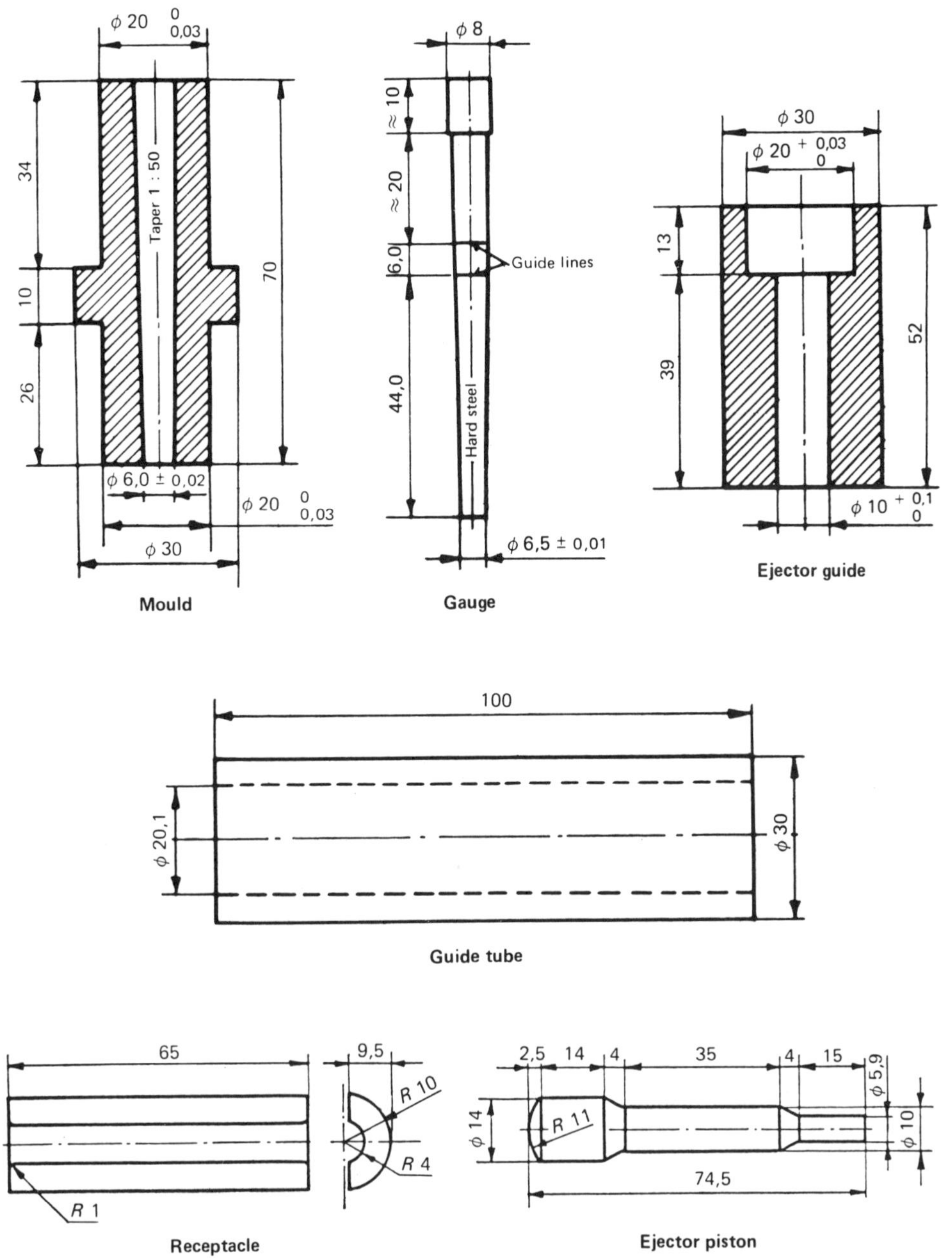

Material :

Mould, ejector guide, guide tube, receptacle – hard bronze.
Gauge, ejector piston – vanadium steel.

FIGURE 2 – **Mould and accessories**

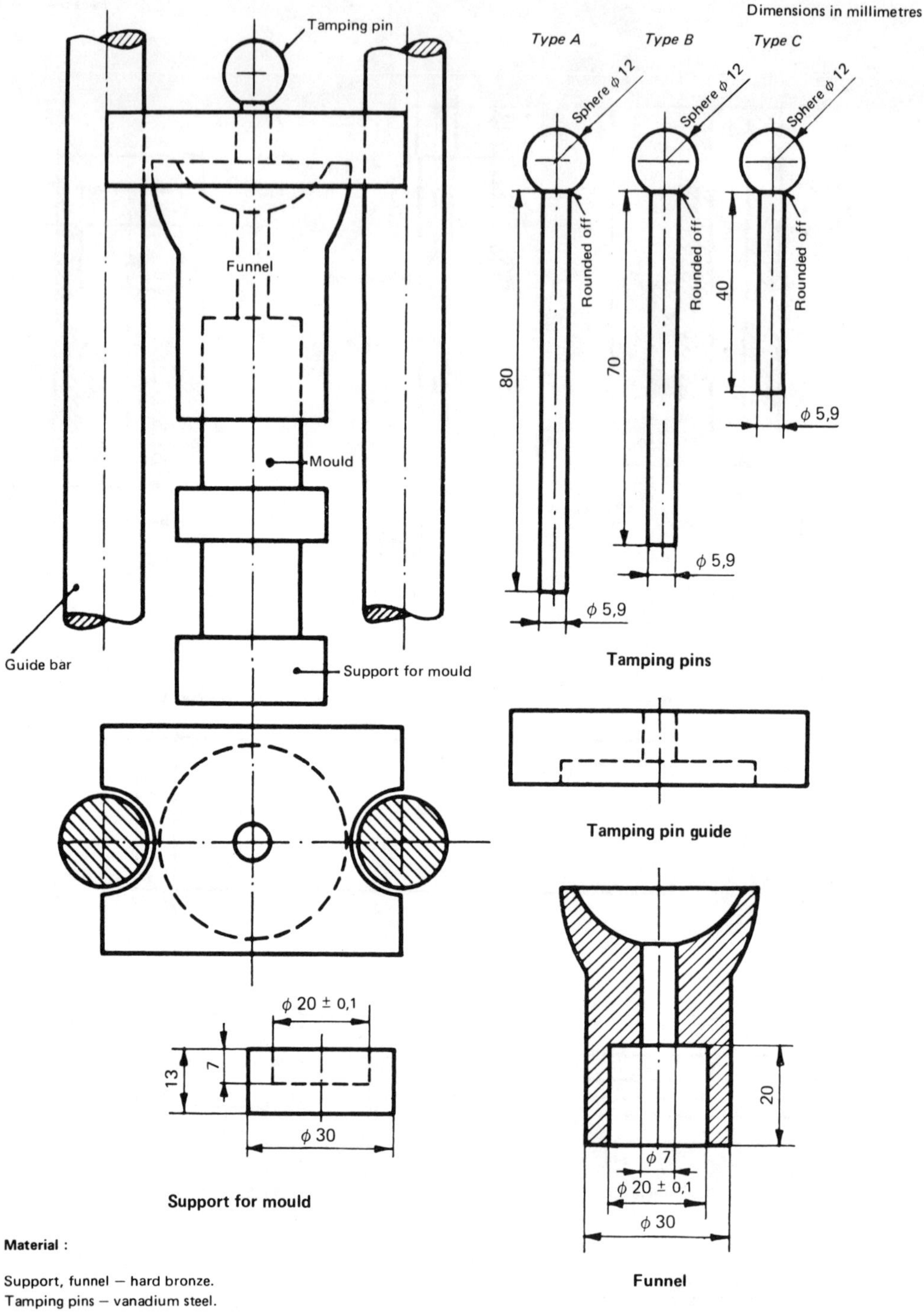

Material :

Support, funnel — hard bronze.
Tamping pins — vanadium steel.

FIGURE 2 A — **Mould and accessories** (*concluded*)

Dimensions in millimetres

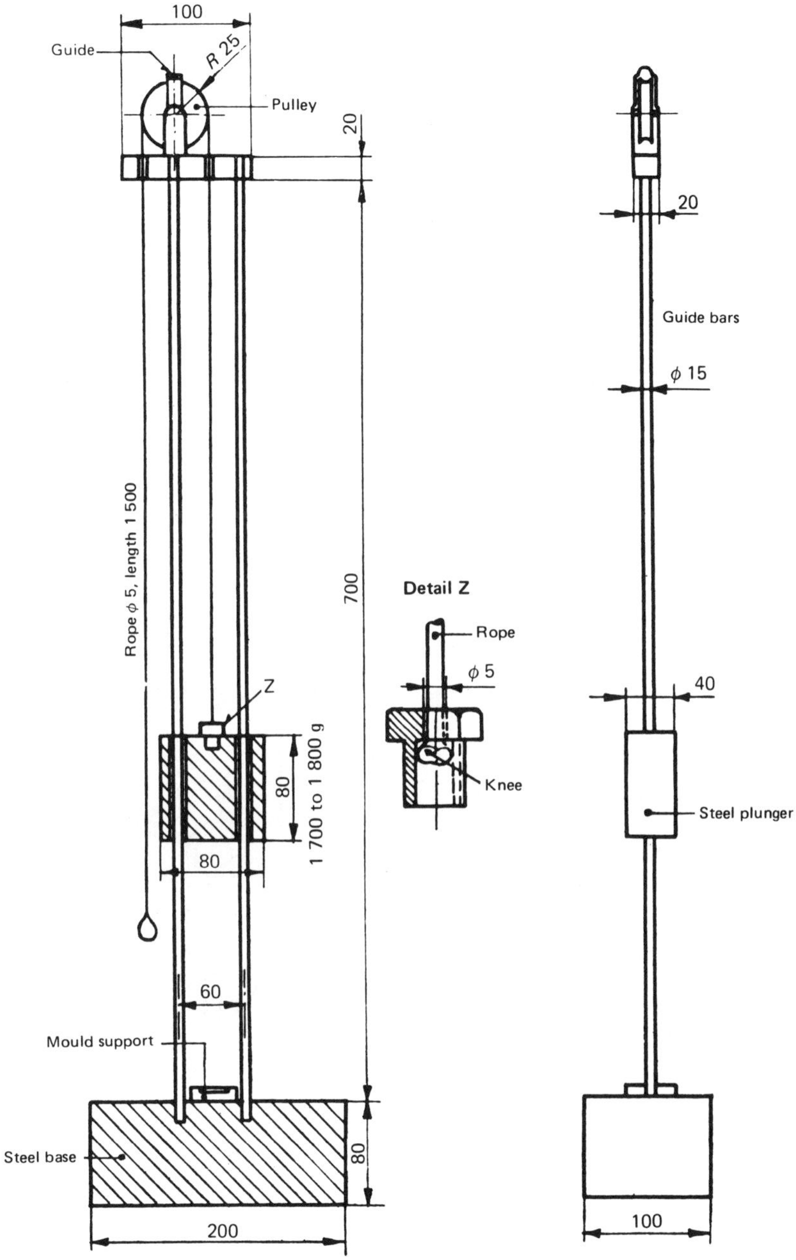

FIGURE 3 – **Ram**

Dimensions in millimetres

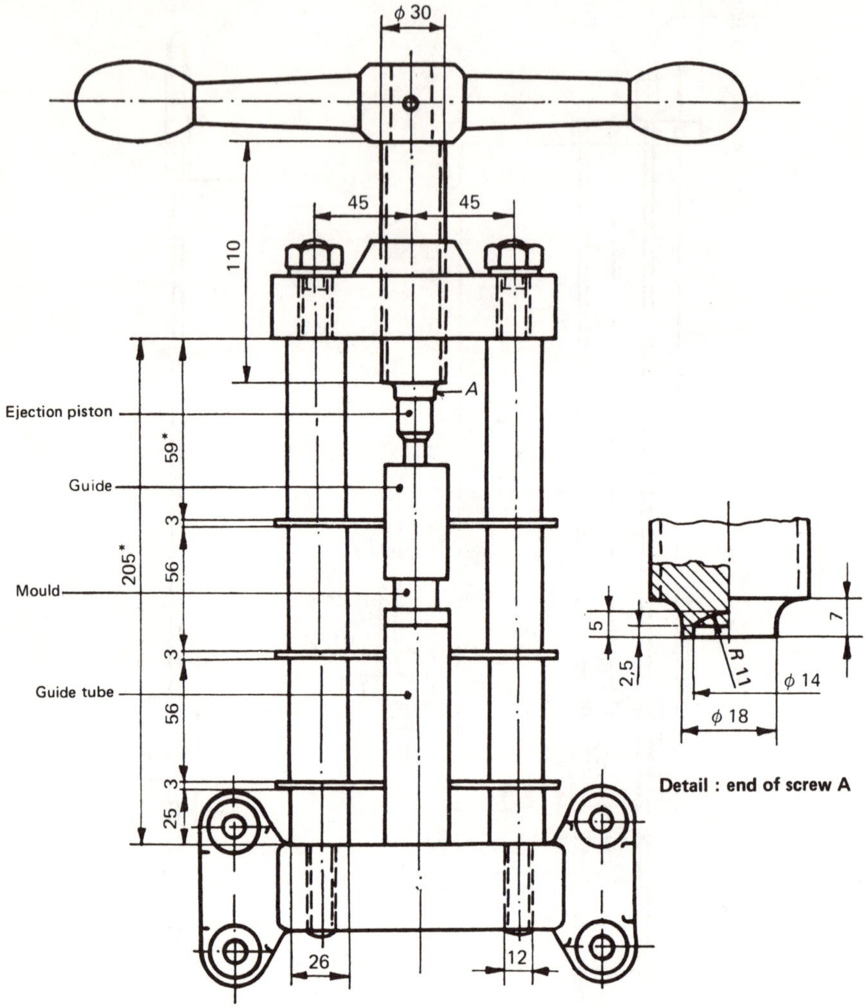

* These dimensions may be increased to 230 and 84 mm respectively.

FIGURE 4 – **Press**

Dimensions in millimetres

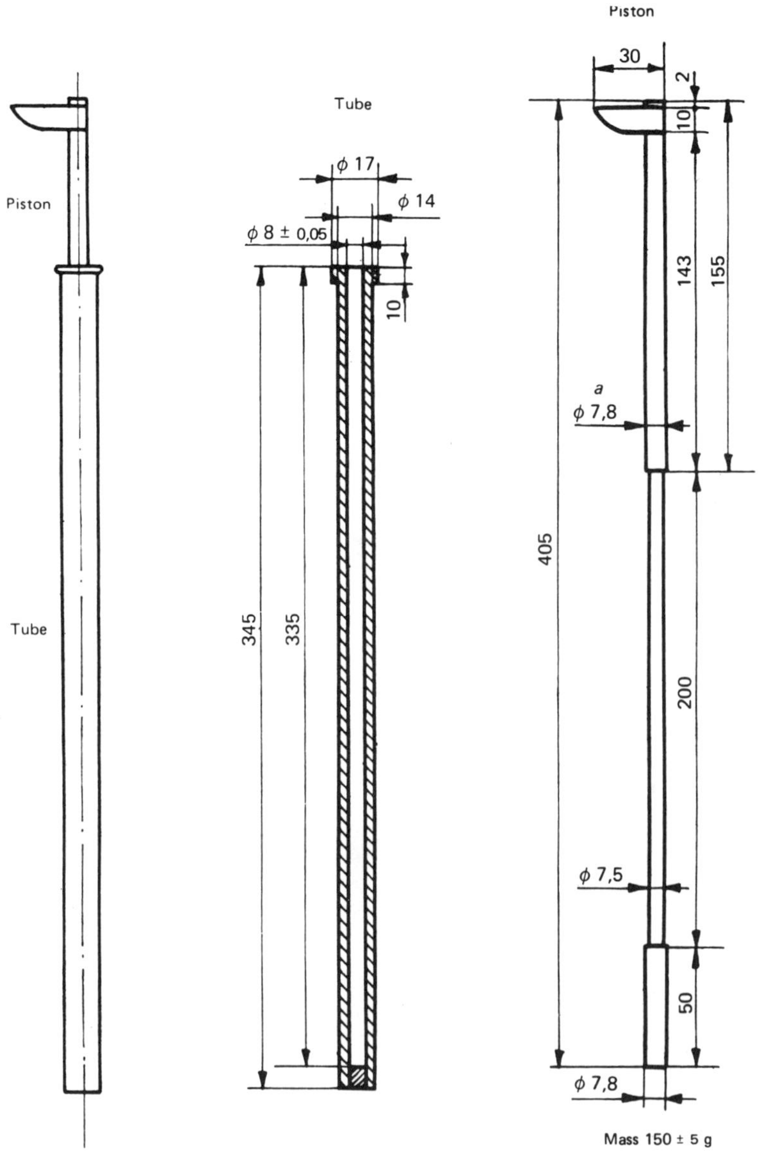

The clearance between tube and piston should be 0,2 ± 0,05 mm

Base welded or screwed and perfectly fluid-tight up to 550 °C

Assembled dilatometer

Material : chrome-nickel steel.

FIGURE 5 – **Dilatometer tube and piston**

Dimensions in millimetres

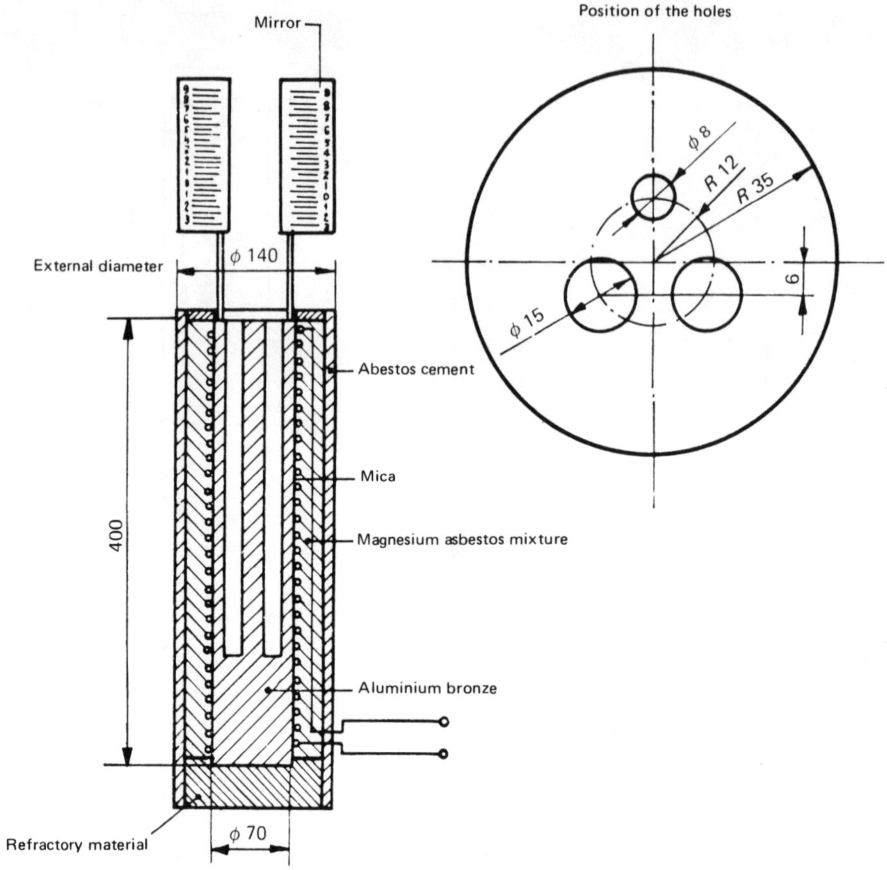

Example : Winding from above downwards (for information)

First	**10** mm,	no winding	
following	**50** mm,	with **18** windings	Wire equivalent to 0,9 mm Kanthal A
following	**300** mm,	with **52** windings	
following	**30** mm,	with **10** windings	
Last	**10** mm,	no winding	

FIGURE 6 – **Example of a suitable electric furnace**

Appendix B

Ruhr Dilatometer

Reprinted with permission from *Recommendations for the dilatometric examination of coal,* Special Publication 17, March 1976, copyrighted by The British Carbonization Research Association, Chesterfield, Derbyshire, Great Britain.

1: GENERAL

1.1: Scope

The International Standard[1] for the classification of hard coals utilizes, among other parameters, a measure of the swelling of coal in a dilatometer. The dilatation of the coal is assessed by the Audibert-Arnu dilatometer[2], a variant of which is the Ruhr dilatometer[3]. Most laboratories in this country make use of equipment based on the design of the Ruhr dilatometer.

The recommended procedures and techniques described herein are those which have been found to permit the greatest reproducibility of results between different laboratories, and will allow comparison with those obtained by the International or German procedures.

The recommendations are based on the use of the Ruhr dilatometer, but no other form of dilatometer is precluded if such apparatus complies with the requirements of this specification.

1.2: Principle

A pencil prepared from powdered coal is heated at a constant rate in a steel retort positioned in a furnace block. The change in level of a piston resting on the sample is observed continuously, and a record is produced which is characteristic of the swelling properties of the coal.

1.3: Definitions

The following parameters, illustrated in Fig. 1, are characteristic:

E: the *softening temperature* (temperature of initial contraction), being that temperature at which the downward movement of the piston is 0.5 mm;

M: the *temperature of maximum contraction*, being that temperature at which the piston reaches its lowest point;

V: the *resolidification temperature* (temperature of maximum dilatation), being that temperature at which the piston reaches its highest point;

k: the *maximum contraction*, the greatest downward movement of the piston, measured from the zero point and expressed as a percentage of the initial pencil length;

d: the *maximum dilatation*, the greatest upward movement of the piston after contraction, measured from the zero point and expressed as a percentage of the initial pencil length.

The following expressions are also used.

Parallel determinations, being tests carried out on two pencils prepared at the same time from a single coal sample, and treated in two retorts in the same furnace during a single heating cycle.

Duplicate determinations, being tests carried out by the same operator using the same furnace, using two pencils prepared from a single coal sample and treated in the same retort in independent furnace heating cycles.

2: APPARATUS

2.1: Mould and accessories

2.1.1: Mould, as in Fig. 2, made from steel, case-hardened after machining. The bore is to be polished after hardening, and the bore and uniformity of taper (i.e. 1 in 50) are to conform to the following dimensions.*

Distance from wide end mm	Bore mm	Tolerance mm
0	7.4	—0.00 + 0.05
10	7.2	
20	7.0	
30	6.8	
40	6.6	
50	6.4	
60	6.2	
70	6.0	

2.1.2: Mallet, plastic head, weight about 200 g.

2.1.3: Ram, as shown in Fig. 2.

2.1.4: Press, as shown in Fig. 2.

2.1.5: Pressure gauge, range 0-3 tonnes, generally as shown in Fig. 2.

2.1.6: Pencil gauge, as shown in Fig. 2.

2.2: Dilatometer

A general arrangement of the dilatometer apparatus is shown in Fig. 3. Critical dimensions with manufacturing tolerances are given in millimetres.

2.2.1: Furnace. The dilatometer furnace shall be of such a form that it is capable of heating two or more retorts of the type described in section 2.2.3 to a temperature of 600°C at a rate of 3 K/min. The acceptability of the heating system shall be assessed in the following manner.

The furnace is heated at a rate of 3 K/min under its normal programmed conditions, and the temperature is observed at the standard sensing point. This is located at a position equivalent to that of the centre of a normally sited pencil and is thus 30 mm above the internal base of a retort. When this temperature has reached about 450°C, a survey is made of the temperature over the lower 250 mm of the retort. The difference between the probe temperature and the mean temperature shown at the normal temperature-sensing position shall be not more than:

±2 K in the lowest 120 mm
±5 K from 120 to 180 mm
±10 K from 180 to 250 mm

For this purpose, temperatures are measured using a suitable prescribed temperature sensor[4,5]. If this is of the platinum resistance type, it should be of a design suitable for point measurement; if of a thermocouple type, it should comprise unsheathed thermocouple elements of diameter not greater than 0.71 mm. The instrument used may either be the recorder specified in section 2.2.2 or another of equal or superior accuracy.

The following apparatus has been found to be generally suitable. The furnace consists of a casing fitted with a base and a top cover. The cover supports in a centre hole a cylindrical block of aluminium bronze (BS 2872, Type CA 103, Condition M is suitable[6]), 65 mm in diameter and 460 mm long. The block has three holes of 380-mm minimum depth and 15-mm diameter, drilled as shown in Fig. 3. The top surface may be insulated by an appropriately shaped piece of board. The block is heated electrically by a resistance winding, suitably insulated, capable of raising the temperature of the block to 600°C at a rate of 3 K/min. The space between the block and the casing is filled with a suitable thermal-insulating material.

For normal test operation a suitable prescribed sensor[4,5] is positioned in a retort situated in the third hole of the block in such a way that the sensor tip lies centrally, 30 mm above the internal base of the retort.

2.2.2: Temperature controller and recorder. It is recommended that the control of temperature should be effected by a separate instrument from that used to record the rise of temperature during test. The controller must be of the automatic programmed type, and should be capable of maintaining a mean rate of temperature rise of 3±0.05 K per minute between 250 and 600°C with a variation of not more than ±1 K per 30-K rise in any 10-minute period. The temperature recorder should be accurate to ±1 K. A

*Suitable gauges for this purpose may be obtained from Messrs J. E. Baty & Co. Ltd., Victoria Road, Burgess Hill, West Sussex.

complete record of the temperature variation during the test shall be obtained.

2.2.3: Retorts and pistons. The arrangement and dimensions of the retorts and pistons are shown in Fig. 3. The dimensions and manufacturing tolerances are given in millimetres.

The retorts are made of cold-drawn seamless tube in heat-resistant steel (BS 970, Type 150 M 19, is suitable[7]) and are fitted with a gas-tight threaded plug at the base and a collar at the top. When inserted in the holes in the furnace block, the retort must be supported only by the collar, with the threaded plug clear of the bottom of the hole. The retort must also be straight, so that it is not restrained by the sides of the hole in the block. It is recommended that the internal bore of each retort is checked with a suitable gauge when new, and again after 100, 150, 200 ... tests. If the internal diameter of the lower third of the retort has increased at any point to more than 8.075 mm, the retort should be rejected. Alternatively, parallel tests should be carried out, comparing the old retort with a new one. Pencils of a given coal prepared in the same mould should give a difference in the values of contraction (k) of not more than 3 units. If there is a consistent difference between the retorts of more than 3 units for a series of four tests, the old retort should be discarded.

The piston is turned from steel rod (BS 970, Type 080 A52, is suitable[7]). The combined weight of the piston and pen assembly should be adjusted to 150 ± 5 g by machining cut-out portions from the piston. The difference between the diameter of the piston and the internal diameter of the retort should be 0.2 ± 0.05 mm on manufacture and this difference should not be allowed to exceed 0.275 mm in use. The piston should slide freely in the retort.

A stand should be provided for the retorts and pistons, to allow them to cool in a vertical position after removal from the furnace. The retorts should be cleaned with an auger, approximately 7.8 mm in diameter and of stem length 400 mm, followed by a wire brush 8 mm in diameter and of stem length 400 mm. A wear-resistant steel reamer, 7.95 mm in diameter and 400 mm in stem length, may also be used.

2.2.4: System for measuring piston movement. Piston movement is recorded on a chart by pens clipped firmly to the tops of the pistons. The chart is affixed to a cylinder rotated uniformly by either clockwork or synchronous motor and mounted on a stand which is clamped to the top of the dilatometer outer casing. Two tests are recorded simultaneously on opposite sides of the chart. The drum diameter and rotation speed are such that a range of 180 K shall occupy not less than 150 mm.

The vertical scale of the chart should be subdivided at 3-mm intervals, corresponding to 5 per cent contraction or expansion of the coal pencil. The scale should be calibrated in units of 50 per cent from —50 to +300.

3: TEMPERATURE CALIBRATION

3.1: Purpose and principle

The measurement of the characteristic temperatures is achieved by the combination of a sensor and its associated indicating or recording instrument. If either the sensor or the instrument is at fault, the observed temperatures will be subject to error.

A method is described (section 3.3) whereby the indicated temperature at two relevant points can be confirmed by direct comparison with the known melting points of samples of two standard reference materials.

3.2: Materials*

(a) Graphite pencils, 30 mm in length, of base diameter 7.4 mm and top diameter 6.8 mm. A small cylindrical depression or reservoir is drilled into the narrow end of each pencil.

(b) Samples of the two standard reference materials:

(1) lead (Analar grade: purity 99.98 per cent) — m.p. 327.0°C

(2) zinc (purity 99.87 per cent) — m.p. 419.3°C

(c) Water-based blacking.

3.3: Procedure

As a precautionary measure, especially when testing zinc, the lower 30 mm of the internal wall, the screw thread of the retort, the sealing plug and the lower face of the piston should be protected from adherent metal by coating with a thin layer of blacking before testing. The blacking, which is water-based, may be applied by any convenient means and should be dried by gentle warming.

Position a lead ball in the recess of the narrow end of a graphite pencil, introduce the pencil into a retort, replace the screw plug and assemble the piston and pen mechanism in the usual manner. Using a zinc ball, a second retort, fitted with its associated piston and pen assembly, is similarly prepared. Insert the retort assemblies in the furnace at approximately 280°C, and proceed as in section 5.2 to evaluate the fusion temperatures of the lead and zinc. (On the chart the melting characteristics, albeit small, are reasonably sharp).

Before re-use, the narrow end of each graphite pencil should be heated in a bunsen flame for a few seconds to permit the molten metal to be shaken from the cylindrical reservoir.

Repeat the procedure, after reversing the disposition of the lead and the zinc in the two retorts.

3.4: Interpretation and significance

(a) It is recommended that a calibration be carried out after 200 tests or after a period of use of 3 months, whichever occurs first.

(b) If the difference between the standard and indicated temperatures is not greater than 7 K, a factor or equation should be established to correct the indicated values.

(c) If the difference between the standard and indicated temperatures is greater than 7 K, malfunction of the indicator or the sensor should be suspected. The accuracy of the temperature-measuring instrument may be confirmed by direct potentiometric calibration against a standard EMF, and the sensor replaced. A sensor calibration should in any case be carried out at much more frequent (e.g. monthly) intervals.

(d) Confirmation of the effect of this corrective action should be achieved by repeating the comparison with the melting points of the two standard reference materials.

*A stock of all these materials is maintained by the British Carbonization Research Association, Chesterfield.

4: PREPARATION OF COAL PENCIL

4.1: Coal sample

4.1.1: Size reduction in one stage. The sample of coal for test is air-dried and reduced in accordance with the procedure[8] described in BS 1017:Part 1:1960 to an upper particle size of 212 μm to yield a laboratory sample of 225 g. The dilatometric examination of the sample shall be commenced immediately after the preparation of the laboratory sample.

4.1.2: Size reduction in two stages. It may be required that the coal be subjected to other analytical techniques at a maximum particle size of 600 μm. Under these conditions, the sample of coal for test is air-dried and reduced by the procedures described[8] to an upper particle size of 600 μm to yield a laboratory sample of 225 g. For the preparation of the dilatometer pencils a 20-g subsample is then crushed to yield a product with a maximum particle size of 212 μm. The final reduction and dilatometric examination are to be commenced as soon as possible after reduction to the 600-μm size.

4.1.3: Final laboratory sample. The crushing operations adopted shall be such that the final laboratory sample shall comply as far as possible with the following size distribution:

μm	%
<212	100
<125	70-60
<63	40-30

4.1.4: Retention of sample. If retention of the sample is required for any reason, the dilatometric properties of a wide range of coals are preserved by storage in sealed glass phials containing an argon atmosphere.

4.2: Preparation of pencil

About 10 g of the ground subsample is well mixed with approximately 1 ml of water by hand-stirring for 2 to 5 minutes.

The mould, with its plug in position and resting on the base holder, is mounted on a firm surface. Approximately 0.5 g of the moistened sample is placed in the mould and the ram placed on top of the coal. The sample is consolidated by three or four sharp blows from the mallet. Further increments, not less than five in all, are added to the mould and similarly consolidated to fill the barrel of the mould. After the last portion has been inserted and consolidated, the pencil is further compressed by application of a load of 1.5 tonnes in the screw press. The application of this load should be continuous and pressure should be released as soon as the required 1.5-t load is indicated by the dial gauge.

The base and plug are removed from the mould barrel. Before the coal pencil is extracted, its wide end is trimmed of irregularities. This is conveniently done by scraping the end of the pencil with a suitable metal straight-edge of an appropriate width to insert into the recess at the base of the mould barrel. A piece of hacksaw blade suitably ground with a square edge is satisfactory for this purpose.

The formed pencil is expelled from the mould by suspending the mould on the carrier arm of the press and screwing the plunger down onto the compressed sample surface. The length of the pencil is reduced to 60±0.5 mm by removing material from the narrow end to conform to the length of a gauge (Fig. 2) 60 mm in length and 8 mm diameter. Any convenient means, such as trimming with a sharp knife, may be used.

Two such pencils are required for the test.

5: PROCEDURE

5.1: Inspection of apparatus

It is essential that the test be carried out with the retorts freely suspended in the furnace block and the dilatometer pistons and retorts scrupulously clean. The cleaning operations should be carried out after each test in the manner outlined in section 5.3.

5.2: Determination of dilatation

Place a prepared coal pencil, narrow end uppermost, in each of two dilatometer retorts and insert a piston into each retort so that it rests on the sample.

Set the furnace controller at a temperature about 75 K below the expected softening temperature of the coal being tested and stabilize the system at this temperature.

For the apparatus described in section 2.2.1, suitable stabilizing temperatures are:

Coal volatile matter % d.a.f.b.	Stabilized dilatometer temperature °C
38	265
28	295
18	325
15	355

Insert the two retorts containing sample and piston in the appropriate holes in the block and allow a 10-minute period for the system to regain the stabilized charging temperature. Attach the pen mechanism to each piston during this period and adjust the pens so that they make contact with the recording chart, approximately at the zero location (a slight offset of the pen from zero is recommended, to facilitate subsequent reading of the chart).

At the end of the 10-minute restabilizing period, commence the heating programme.

After approximately 10 minutes, start the recorder drum and index the furnace temperature, as indicated by the recorder, on the drum chart.

The conclusion of the test is denoted when the pen has traced a horizontal line for 5 minutes after completion of the dilatation process, or when no dilatation has occurred up to 500°C. In the 5-minutes interval after completion of the dilatation process, or at about 500°C when no dilatation has occurred, a second indexing of furnace temperature is made from the recorder on the drum chart. If, in the period between the indexed temperatures on the drum chart at the beginning and the end of the test, the recorder shows a rate of temperature rise differing from 3 K/min, i.e. 30±1K in any 10-minute interval, the test shall be deemed invalid (but see section 2.2.2).

At the end of the test, the pen mechanisms are detached and the retort and piston assemblies withdrawn from the furnace. The pistons may be removed from the retorts and independently cooled by suspending in air.

5.3: Cleaning of the furnace, retort and piston

The following method of checking and cleaning is recommended.

Furnace. Check that each retort is capable of being freely suspended from the collar. If not, clean out the holes in the furnace block.

Retort and plug. Remove the plug, crush the semicoke and remove as much of it as possible with the auger. Complete the cleaning of the retort by use of the wire brush, and also with the reamer if required. Clean the plug end with emery paper (not coarser than grit no. 220).

Piston. Clean the piston, including its base, with wire wool and emery paper (not coarser than grit no. 220), taking care not to round the edges, and check that the piston slides freely in the cleaned retort. (For the tolerance between the piston and the retort, see section 2.2.3).

6: REPORTING OF RESULTS

Five basic parameters are obtained from the charts after completion of a test, as defined in section 1.3 and illustrated in Figs 1 and 4. These are:

Softening temperature, *E*
Temperature of maximum contraction, *M*
Resolidification temperature, *V*
Maximum contraction, *k*
Maximum dilatation, *d*

The dilatation is positive if the final line of maximum dilatation is above the zero line (Fig. 4i), and negative if below it (Fig. 4ii). If the dilatometer curve does not rise after the initial contraction, the dilatation behaviour is noted as "contraction only" (Fig. 4iii). If the final trace of the curve is not truly horizontal, but slopes downward (Fig. 4iv), it is conventional to report the contraction as that value observed at 500°C.

If the maximum dilatation is greater than 300 per cent it shall be reported as "d>300".

The results of acceptable replicate determinations (see section 7) shall be averaged; the temperature parameters, the contraction and the dilatation shall each be rounded to the nearer whole number.

7: TOLERANCES

7.1: Application to dilatometry

The techniques and procedures involved in the dilatometric examination of coal are such that determinations may be carried out as parallel tests or duplicate tests (see section 1.3). The use of parallel tests has been recommended, but duplicate determinations may also by made. For the purposes of this section, the tolerances given for determinations made within one laboratory shall be taken to refer to either parallel tests or duplicate tests.

7.2: Definitions

The permitted tolerances are described by the terms "repeatability" and "reproducibility", which are defined as follows.

Repeatability (within-laboratory): the maximum acceptable difference between parallel determinations carried out simultaneously, or between duplicate determinations carried out at different times, in the same laboratory on pencils prepared from the same crushed laboratory sample by one operator using the same apparatus. Differences greater than the stated tolerance may be accepted after further replicate testing (see section 7.4).

Reproducibility (between laboratories): the maximum acceptable difference between the mean of acceptable replicate determinations carried out in one laboratory and the mean of acceptable replicate determinations carried out in any other laboratory on pencils prepared from representative samples taken from the same bulk sample after reduction to pass a test sieve of aperture not greater than 4.0 mm.

7.3: Tolerances

	Repeatability (within-laboratory)	Reproducibility (between laboratories)
Temperature parameters	7 K	15 K
Contraction, k	5 units	8 units
Dilatation, d If d =		
0-50	5	10
51-100	10	15
101-200	15	20
>200	20	25

7.4: Confirmatory (repeat) tests

7.4.1: Duplicate or parallel determinations. If the results obtained exceed the given repeatability limit, two further determinations shall be carried out immediately, and the following procedure shall be adopted.

(a) If the spread of the four results is less than $1.3T$, where T is the given repeatability, all four results shall be accepted, and the mean value reported.

(b) If the spread is greater than $1.3T$ but three of the results lie within $1.2T$, these three shall be accepted and the fourth rejected.

(c) If the results do not satisfy either of these conditions, an investigation should be made of all the circumstances involving the sample preparation, the apparatus and the method of examination.

7.4.2: Determinations by different laboratories. Reported results from different laboratories with differences less than the given reproducibility limits are all equally valid. If differences persistently greater than these limits are observed, the need for further investigation into the methods of sampling and testing is indicated.

REFERENCES

1. United Nations Economic Commission for Europe. International classification of hard coals by type. Geneva, 1956.
2. International Organization for Standardization. Hard coal: Audibert-Arnu dilatometer test. ISO Standard 349, 1975.
3. Deutscher Normenausschuss. Testing of solid fuels: determination of the coking capacity of bituminous coal by the dilatometer. DIN 51 739, 1976.
4. British Standards Institution. Code for temperature measurement. Industrial resistance thermometry. BS 1041 : Part 3 : 1969.
5. British Standards Institution. Code for temperature measurement. Thermocouples. BS 1041 : Part 4 : 1966.
6. British Standards Institution. Copper and copper alloys. Forging stock and forgings. BS 2872 : 1969.
7. British Standards Institution. Wrought steels in the form of blooms, billets, bars and forgings. Carbon and carbon manganese steels including free cutting steels. BS 970 : Part 1 : 1972.
8. British Standards Institution. The sampling of coal and coke. Sampling of coal. BS 1017 : Part 1 : 1960.

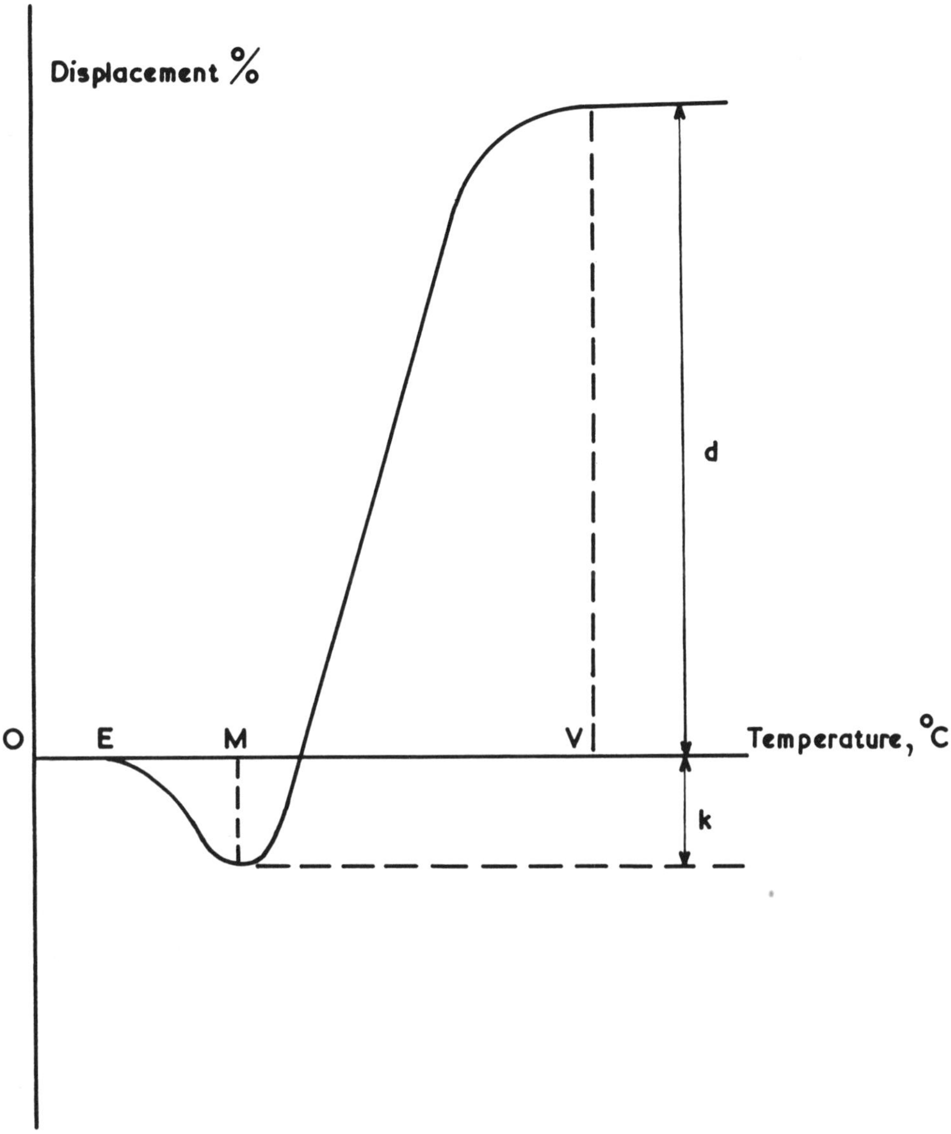

Fig. 1. Characteristic dilatation curve

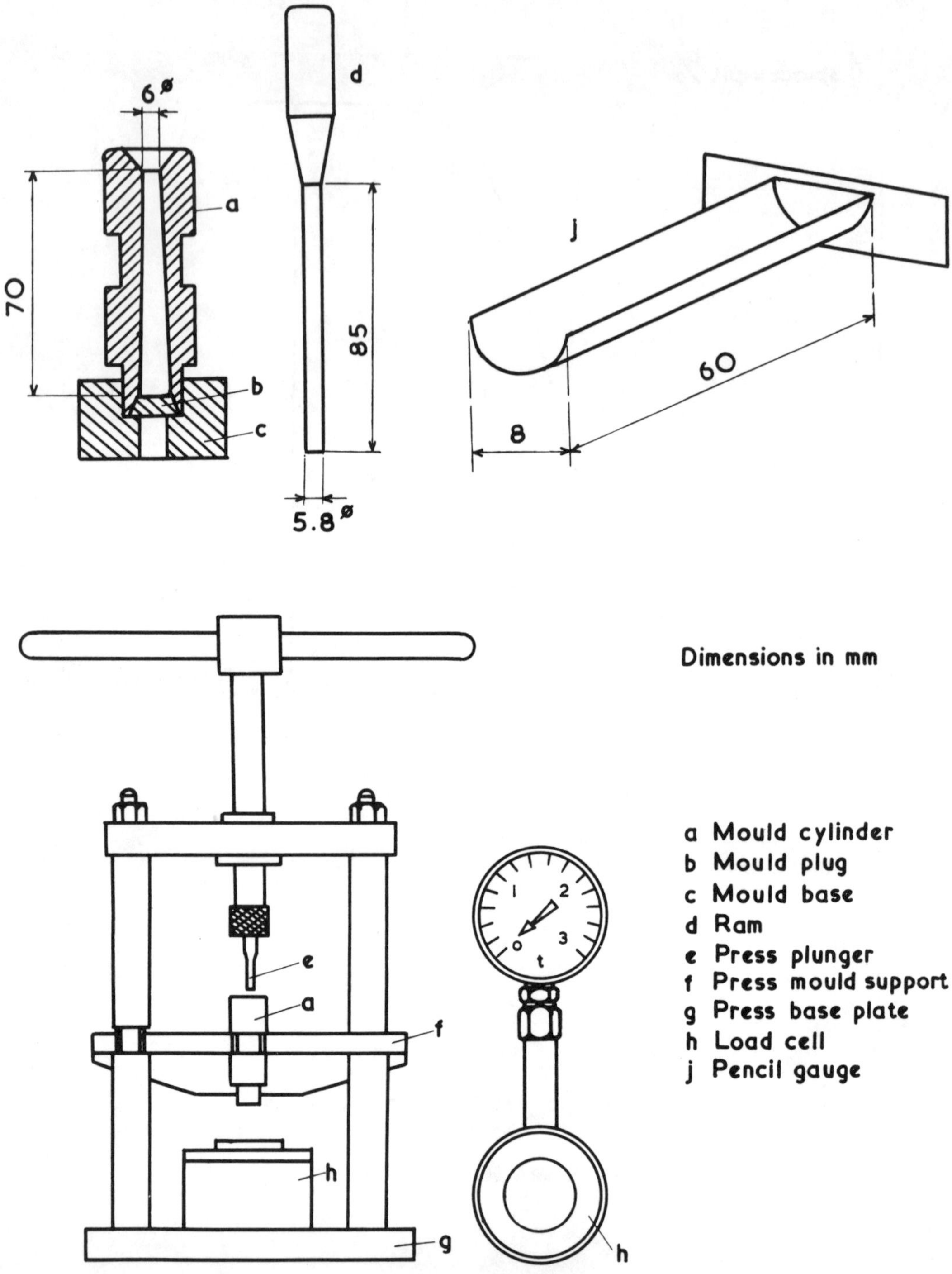

Fig. 2. Mould, press and accessories

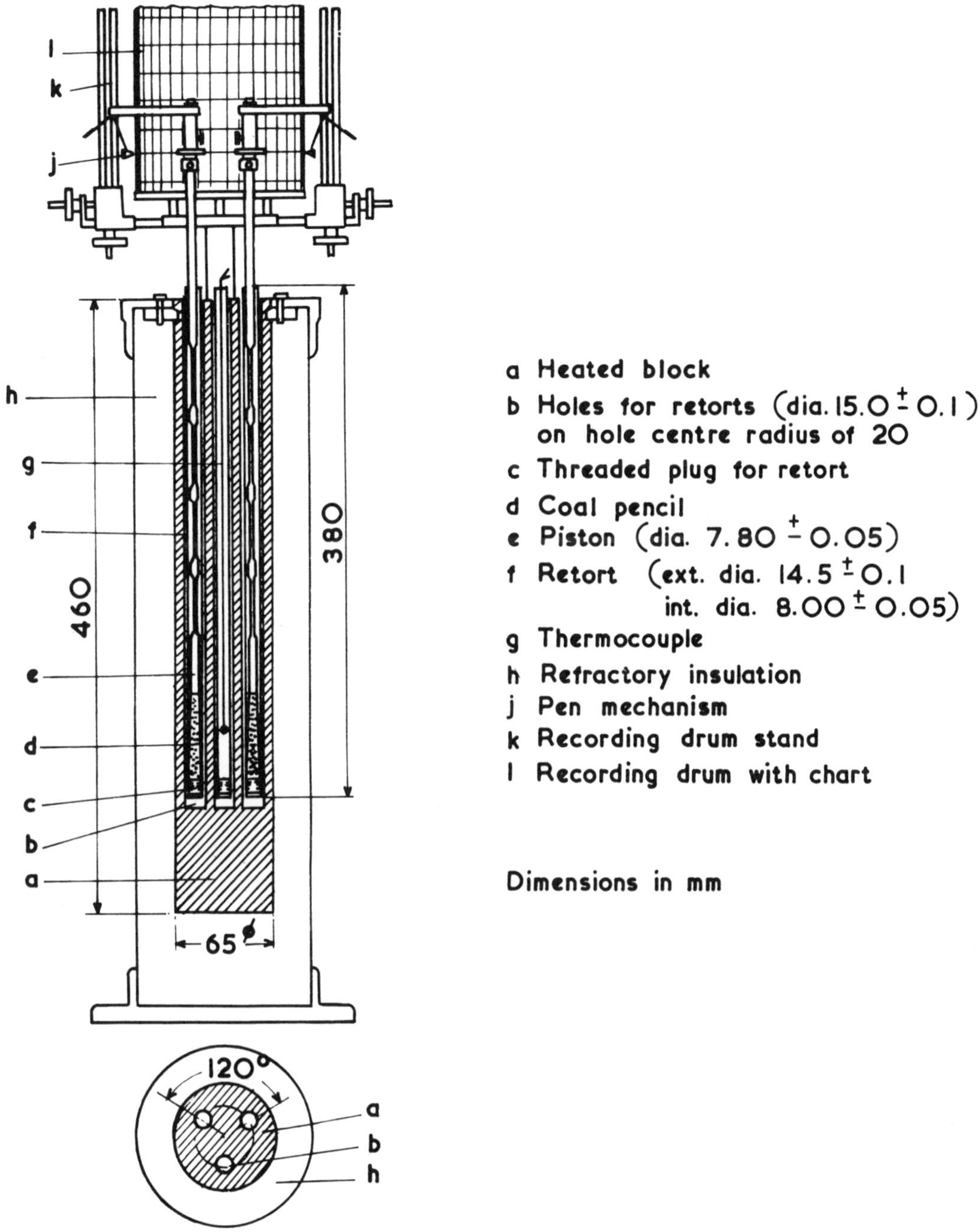

Fig. 3. Dilatometer apparatus

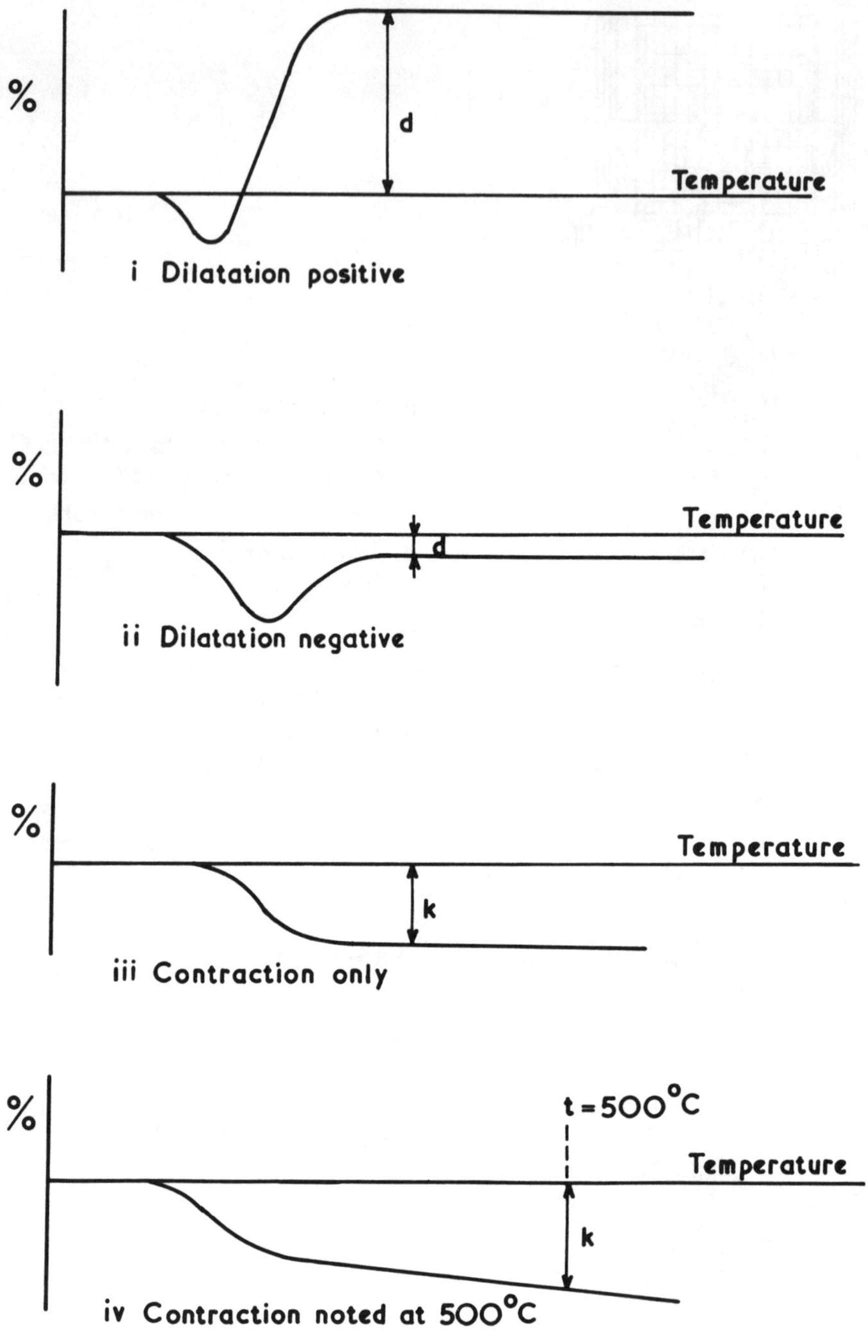

Fig. 4. Types of curve

Appendix C

Coal—Determination of Caking Power— Gray-King Coke Test
ISO Standard 502-1974 (E)

This material is reproduced with permission from the International Organization for Standardization, Standard 502-1974(E) "Coal—Determination of Caking Power—Gray-King Coke Test," copyrighted by the American National Standards Institute, 1430 Broadway, New York, New York 10018.

Coal – Determination of caking power – Gray-King coke test

0 INTRODUCTION

The purpose of the Gray-King coke test, which is one of the parameters adopted for the International Classification of Hard Coal by Type by the United Nations Economic Commission for Europe, is to assess the caking properties of a type of coal or a blend of coals by carbonizing under standard conditions.

Although the Gray-King test and the Roga test both assess the caking properties of a coal they do not measure precisely the same parameters and should not be regarded as alternative methods.

1 SCOPE AND FIELD OF APPLICATION

This International Standard specifies a method of assessing the caking power of coal under standard conditions.

2 PRINCIPLE

The coal is heated under standard conditions to a final temperature of 600 °C. The coke residue obtained is classified by reference to a series of standard residues (see figure 1). If the coke residue produced is so swollen that it fills the cross-section of the retort tube, the determination is repeated with the coal admixed with a suitable quantity of electrode carbon or equivalent material (see 9.1). For these highly swelling coals the Gray-King coke type is defined by the minimum amount of electrode carbon required to produce a strong hard coke residue of the same volume as the original coal and electrode carbon mixture.

3 REAGENT

Standard electrode carbon (see 9.1)

High temperature electrode carbon :

Moisture	less than 1 %
Volatile matter	less than 1,5 %
Ash	less than 5 %
Bulk density at 25 °C (see the annex)	1,00 to 1,05 g/cm^3
Relative density at 25 °C	2,05 to 2,09 (see 9.2)

Size analysis :

Retained on 212 μm test sieve	nil
Through 212 μm test sieve, retained on 125 μm test sieve	less than 26 %
Through 125 μm test sieve, retained on 63 μm test sieve	10 to 40 %
Through 63 μm test sieve	50 to 85 %

4 APPARATUS

4.1 Furnace

Horizontal electric furnace, 50 mm internal diameter and 300 mm long, with one end closed and the other carrying a plug of insulating material which is bored centrally with a hole 25 mm in diameter. The winding of the furnace shall be such that the middle 200 mm is at a uniform temperature within ± 5 °C at both 300 and 600 °C. Alternatively the furnace may be constructed from an electrically-heated aluminium-bronze block, with one or several bores of 25 mm diameter. The furnace should be insulated and located in a cover of metal or other suitable material, and should be equipped with a suitable thermocouple, lying above the retort tube when the latter is in position and with the junction at the centre of the furnace. An indicator should be provided for showing the furnace temperature with an accuracy of ± 5 °C. A suitable means of controlling the energy input should also be provided to permit an increase in temperature at a rate of 5 °C per minute. A multiple tube furnace to allow simultaneous determinations is convenient. The furnace may be of the fixed type or mounted on rails. Suitable furnaces are shown in figures 2 and 3.

4.2 Retort tube (figure 4)

A heat-resistant glass or transparent silica tube, 20 mm internal diameter and 300 mm long, closed at one end, with a side arm, 8 mm internal diameter and 50 mm long, sealed in at a distance of about 20 mm from the open end. The tube shall be smooth and either of uniform bore, or with a slight taper (19 mm to 21 mm) such that the open end is the larger.

4.3 Distance rod

A rod with a flat disk at one end to assist in the packing of the coal and to indicate the free end of the coal sample in the retort tube.

4.4 Receiver and outlet tube

A glass vessel of adequate size suitably supported and attached to the side arm of the retort tube, fitted with an outlet tube leading to atmosphere or to a piece of small bore silica tubing at the end of which the gas leaving the receiver can be burned. The receiver may conveniently be a U-tube which can be immersed in water.

5 PREPARATION OF SAMPLE

The coal used for the determination of the Gray-King coke type is the analysis sample ground to pass a sieve of 0,2 mm aperture. If necessary, expose the sample in a thin layer for the minimum time required for the moisture content to reach approximate equilibrium with the laboratory atmosphere.

Before commencing the determination mix the air-dried sample of coal thoroughly for at least 1 min, preferably by mechanical means. The sample shall be prepared on the same day as the determination is carried out.

6 PROCEDURE

6.1 Coals with a Gray-King coke type within the range A to G_2 (see 9.3)

Raise the temperature of the furnace until it is steady at 325 °C.

Weigh on a scoop 20 to 20,01 g of the coal sample and transfer it to the retort tube, held in such a manner that the coal cannot enter the side arm. Complete the transfer with a soft brush and allow the coal to fall to the far end of the retort tube. Hold the tube horizontally, insert the distance rod so that the face of the disk is 150 mm from the closed end of the retort tube and spread the coal into a layer of uniform depth by shaking and turning. Withdraw the distance rod and insert a flattened pad of asbestos wool or a notched asbestos disk to retain the coal in position. Without disturbing the position of the coal, close the open end of the retort tube with a heat-resisting stopper. Connect the receiver to the side arm and insert the retort tube in position in the furnace so that the centre of the coal layer coincides with the centre of the furnace. If the furnace is mounted on rails, clamp the retort tube in a horizontal position and run the furnace into position.

Raise the energy input to the furnace in such a manner that the temperature of 325 °C is regained in 3 to 7 min and maintain a uniform rate of rise of 5 °C per minute thereafter until a temperature of 590 °C is reached. At this point, regulate the energy input to the furnace so that a temperature of 600 °C is reached, and maintain this temperature for 15 min.

Withdraw the retort tube (or retract the furnace) and allow it to cool. Detach the receiver, remove the stopper and slide the coke residue out for examination.

6.2 Coals with a Gray-King coke type greater than G_2 (see 9.3)

Weigh X g of the electrode carbon (3.1), where X is always an integer, into a weighing bottle and add $(20 - X)$ g of the coal sample. Insert the stopper and mix the contents thoroughly.

Transfer the mixture to the retort tube and proceed exactly as described in 6.1.

Repeat the determination if necessary, varying the amount of electrode carbon in 20 g of the mixture, until a coke residue of type G is obtained using the minimum mass of electrode carbon.

7 EXPRESSION OF RESULTS

Report the Gray-King coke type of a coal by reference to figure 1 and the table on page 5, where the appearance of typical coke residues is described and illustrated. For coals giving a coke type with an index greater than G_2, the subscript defines the minimum number of grams of electrode carbon added to produce a standard G type coke residue.

8 PRECISION OF THE METHOD

Coke type	Repeatability	Reproducibility
A to G_1	One letter	One letter
Greater than G_1	One unit in the subscript	One unit in the subscript

8.1 Repeatability

The results of duplicate determinations, carried out at different times in the same laboratory by the same operator with the same apparatus on the same analysis sample, shall not differ by more than the above value.

8.2 Reproducibility

The means of the results of duplicate determinations, carried out in each of two different laboratories on representative portions taken from the same analysis sample after the last stage of sample preparation, shall not differ by more than the above value.

9 NOTES ON PROCEDURE

9.1 It has been shown that materials other than standard electrode carbon may be used, although it has not yet been possible to define parameters to include all such materials. Any material which has been found by experiment to produce results equivalent to those obtained when using standard electrode carbon may be used. To enable these experiments to be carried out, the United Kingdom is willing to supply a sample of equivalent anthracite with the following size range :

Retained on 212 μm test sieve	nil
Through 212 μm, retained on 125 μm	5 to 10 %
Through 125 μm, retained on 63 μm	20 to 25 %
Through 63 μm	65 to 75 %

9.2 Determine relative density by the normal method using the density bottle. To ensure complete wetting of the electrode carbon, use a 1 % solution of a wetting agent and evacuate the density bottle containing the electrode carbon and wetting agent solution to a pressure of 80 mbar in a vacuum desiccator. Maintain this vacuum for 10 min before transferring the density bottle to a water-bath thermostatically controlled at 25 °C.

9.3 Although there is no precise relationship between the crucible swelling number and the Gray-King coke type, the following table shows the broad relationship to be expected. This will be found useful for indicating the necessity for blending with electrode carbon and the probable amount required.

Crucible swelling number	Gray-King coke type
0 to 1/2	A to B
1 to 4	C to G_2
4 1/2 to 6	F to G_4
6 1/2 to 8	G_3 to G_9
8 1/2 to 9	G_7 or above

The above relationship is known to be applicable to United Kingdom coals and is intended only as a general guide. Each country should determine the correlation applicable to its own coals.

10 TEST REPORT

The test report shall include the following particulars :

a) the reference of the method used;

b) the results and the method of expression used;

c) any unusual features noted during the determination;

d) any operation not included in this International Standard, or regarded as optional.

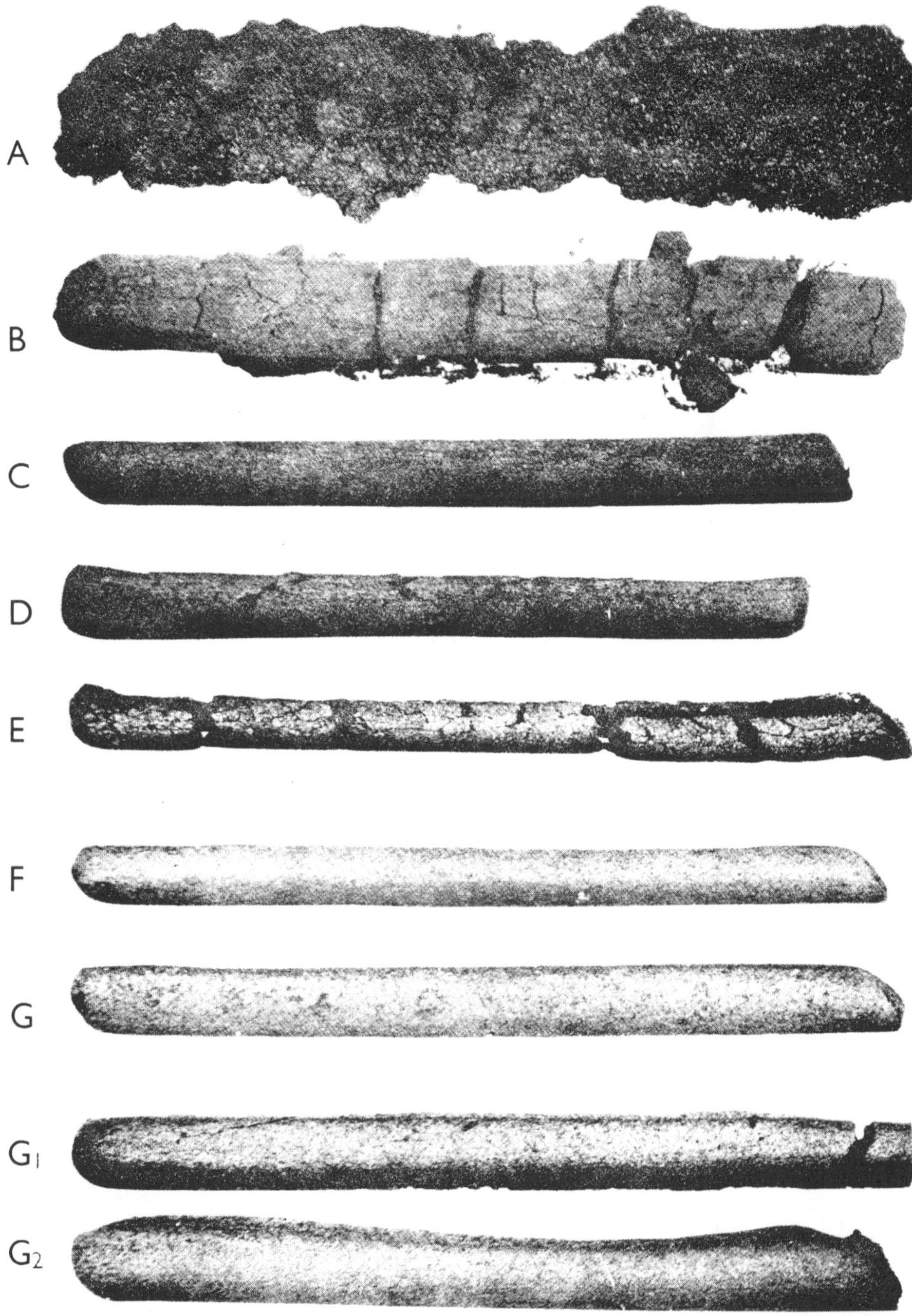

FIGURE 1 – Gray-King coke type

COPYRIGHT RESERVED

Published with permission of The Controller of Her Majesty's Stationery Office, London, U.K.

TABLE — **Classification of Gray-King coke type**

A, B and C			D, E and F			G	G_1 to G_x		
Retains initial cross-section			shrunken			retains initial volume	swollen		
Examine for strength			Examine for strength			Examine for strength	Examine for degree of swelling		
NON-COHERENT	**BARELY COHERENT**	**COHERENT**	**MODERATELY HARD AND SHRUNKEN**	**HARD AND VERY SHRUNKEN**	**HARD, STRONG AND SHRUNKEN**	**HARD AND STRONG**	**SLIGHTLY SWOLLEN**	**MODERATELY SWOLLEN**	**HIGHLY SWOLLEN**
Usually in powder form but may contain some pieces which, however, cannot be handled without breaking	In several pieces and some loose powder. Pieces can be picked up but break into powder on handling	Usually in one piece but easily broken; may be in two or three pieces with practically no loose powder; very friable and dull	May be fissured but can be scratched with fingernail and stains the fingers on rubbing the curved surface vigorously; usually dull and black and appearing fritted rather than fused	Usually very fissured; moderate metallic ring; does not stain the fingers on rubbing; grey or black with slight lustre	May be fissured; moderate metallic ring; does not stain the fingers on rubbing. Cross-section well fused and greyish	Well fused with a good metallic ring when tapped on a hard wooden surface			G_3 and higher Guided by swelling number, blend with minimum number of parts of electrode carbon to give a standard G-type coke
A	**B**	**C**	**D**	**E**	**F**	**G**	G_1	G_2	G_x

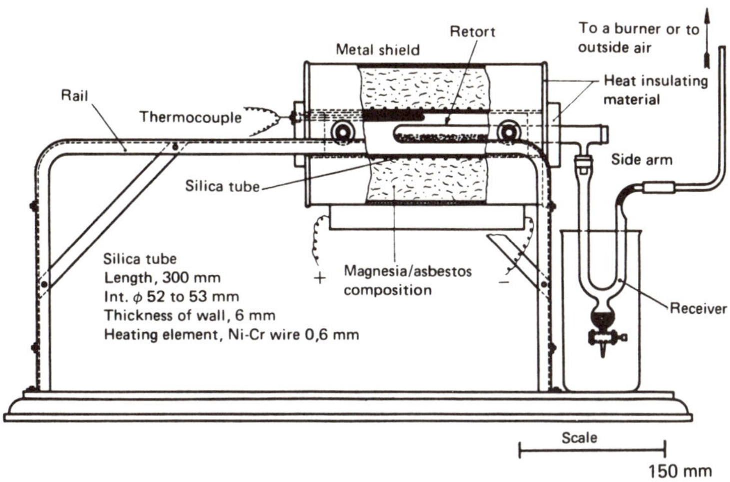

FIGURE 2 – **Single tube furnace**

ANNEX

DETERMINATION OF THE BULK DENSITY OF ELECTRODE CARBON

A.1 APPARATUS

A.1.1 Dropping box

A box as shown in figure 5, screwed to a rigid bench or table. The pad at the base of the box shall have a hardness of 71 to 80 International Rubber Hardness Degrees.

Dimensions in millimetres

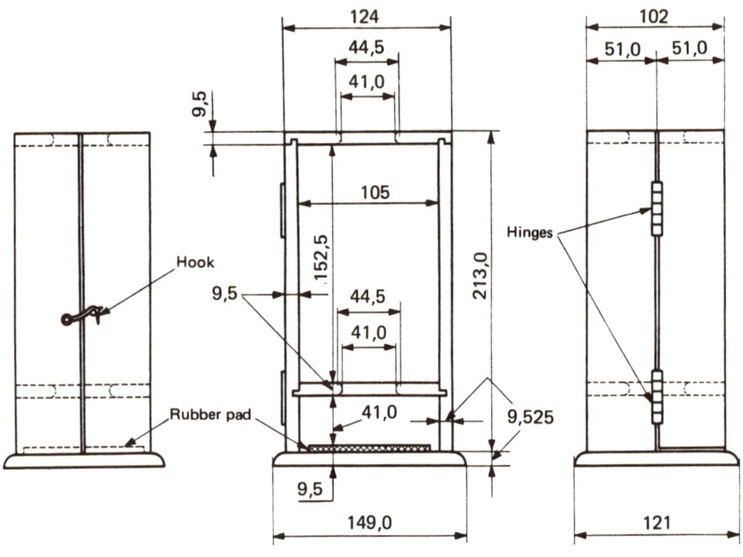

FIGURE 5 – **Dropping box**

Appendix D

Hard Coal—Determination of Caking Power—Roga Test
ISO Standard 335-1974 (E)

This material is reproduced with permission from the International Organization for Standardization, Standard 335-1974(E) "Hard Coal—Determination of Caking Power—Roga Test," copyrighted by the American National Standards Institute, 1430 Broadway, New York, New York 10018.

Hard coal – Determination of caking power – Roga test

0 INTRODUCTION

The purpose of the Roga test, which provides one of the parameters adopted for the "International Classification of Hard Coal by Type" of the United Nations Economic Commission for Europe, is to assess the caking power of a coal under standard conditions.

Although the Roga test and the Gray-King test both assess the caking properties of a coal, they do not measure precisely the same parameters and shall not therefore be regarded as alternative methods.

1 SCOPE AND FIELD OF APPLICATION

This International Standard specifies a method of determining the caking power of hard coal by the Roga test.

2 PRINCIPLE

The caking power of a coal is defined by the mechanical strength of the crucible coke obtained by carbonization under standard conditions of an intimate mixture of 1 g of coal and 5 g of standard anthracite. The crucible coke obtained is then submitted to a drum test in a strictly prescribed manner and the Roga Index is calculated from the results.

3 REAGENT

3.1 Standard anthracite

An anthracite of E.C.E. Code No. 100 A, having an ash content of less than 4 % on the dry basis and a volatile matter number of 5 to 6,5 % on the dry, ash-free basis (see note). Carefully sieve the anthracite before use on sieves of 0,3 mm and 0,4 mm apertures (square); retain the fraction passing the 0,4 mm sieve and remaining on the 0,3 mm sieve for the test.

NOTE – An anthracite of E.C.E. Code No. 100 A but having a different ash and volatile matter content may be used, provided it has been found by experiment to give the same results, within the tolerances of the method (see clause 8), as those obtained using the standard anthracite.

4 APPARATUS

The balance shall be sensitive to 0,01 g.

4.1 Crucible, porcelain, having the following dimensions :

external diameter at the top	40 ± 1,5 mm
external diameter at the base	20 ± 1,5 mm
external height	40 ± 1,5 mm
wall thickness	less than 2 mm

4.2 Lid : heat-resistant steel plate, 55 mm square, 1,5 to 2 mm thick, with a hole 2 mm in diameter in the centre.

4.3 Stirrer, made of wire, having an 8 mm loop at one end.

4.4 Heat-resistant steel weight, of 110 to 115 g, of the following dimensions : 31 mm in diameter and 21 mm in height, with a threaded hole in the centre 7 mm in diameter and 10 to 12 mm deep into which can be screwed a suitable rod for lifting the weight from the crucible.

4.5 Press for compressing the mixture of coal and standard anthracite under a weight of mass 6 kg (see figure 1).

4.6 Electric furnace, with a zone of uniform temperature and a temperature control device capable of maintaining that zone at 850 ± 10 °C.

4.7 Drum with a cover, driving shaft, transmission gear and electric motor for carrying out the abrasion test on coke. The drum (see figure 2) has a 200 mm internal diameter, is 70 mm deep and is made of 2 mm thick sheet iron. To the inside walls, which shall be perfectly finished, are welded two symmetrical sheet iron strips 70 mm long, 30 mm wide and 2 mm thick. To close the drum, the cover is seated on a felt or rubber gasket and is secured by four wing-nuts. The drum is fitted with a threaded stub axle which serves to couple it to the driving shaft. It is convenient to mount two drums on a single shaft to enable two tests on two different samples to be carried out simultaneously. The drum is rotated with the stub axle horizontal at 50 ± 2 rev/min.

4.8 Laboratory sieve, made of thin brass sheet, with 1 mm round holes.

4.9 Stopwatch.

4.10 Asbestos sheet.

4.11 Brush.

5 PREPARATION OF SAMPLE

Crush the sample of air-dried coal to pass a 0,2 mm test sieve. Take care to avoid an excessive production of particles smaller than 0,1 mm. It is essential that not less than 40 % of the total sample consists of particles between 0,1 and 0,2 mm. Mix the crushed sample thoroughly for at least 1 min, preferably by mechanical means.

6 PROCEDURE

It is emphasized that duplicate determinations are to be carried out on each sample of coal, but that these shall not be done simultaneously.

Weigh a clean, dry crucible, place in it 1 g of the coal and 5 g of the standard anthracite (3.1), both weighed to an accuracy of 0,01 g. Mix carefully for 2 min with the stirrer (4.3), level off the surface and place the steel weight on it. Press the whole for at least 30 s under a weight with a mass of 6 kg. Remove the crucible from the press and cover it with the lid, leaving the steel weight in the crucible.

Raise the temperature of the furnace to 850 ± 10 °C and insert the crucible (see note below). Check the temperature on the floor of the furnace at the side of the crucible and ensure that the temperature of 850 ± 10 °C is regained within 5 min from the insertion of the crucible. After a total heating time of 15 min, remove the crucible from the furnace and allow it to cool on an asbestos plate for 45 min.

After cooling, remove the weight from the crucible by means of the rod. Brush back into the crucible any particles of coke adhering to the weight and weigh the crucible plus contents to an accuracy of 0,01 g. Transfer the contents of the crucible very carefully to the sieve. Carefully transfer back to the crucible the larger pieces of coke, using a pair of tweezers. Sieve the remaining coke very carefully and transfer the oversize back to the crucible. Reweigh the crucible plus contents. Transfer the contents of the crucible to the drum and fit the cover. Connect the drum to the axle, start the stopwatch and rotate the drum for 5 min at a speed of 50 ± 2 rev/min. Remove the coke from the drum and sieve it again. Transfer the coke remaining on the sieve back to the crucible and reweigh. Return the coke from the crucible to the drum and repeat the abrasion procedure, seiving and reweighing of the oversize exactly as described above. Carry out a third abrasion test under the same conditions, sieve and weigh the final oversize. All weighings shall be made to an accuracy of 0,01 g.

NOTE — For convenience, a second crucible may be heated in the furnace at the same time; it is emphasized, however, that duplicate determinations on a single sample of coal shall not be carried out simultaneously.

7 EXPRESSION OF RESULTS

The Roga Index (R.I.) is given by the formula

$$\frac{100}{3\,m_1}\left(\frac{m_2 + m_5}{2} + m_3 + m_4\right)$$

where

m_1 is the total mass, in grams, of crucible coke after carbonization;

m_2 is the mass, in grams, of coke remaining on the sieve before the first abrasion test;

m_3 is the mass, in grams, of coke remaining on the sieve after the first abrasion test;

m_4 is the mass, in grams, of coke remaining on the sieve after the second abrasion test;

m_5 is the mass, in grams, of coke remaining on the sieve after the third abrasion test.

The Index shall be reported to the nearest integer, as the mean of the results of two separate determinations which do not differ by more than 3 units (see clause 8). Where the difference is greater, the test shall be repeated.

8 PRECISION OF DETERMINATION

Caking power	Maximum acceptable differences between results	
	Same laboratory (Repeatability)	Different laboratories (Reproducibility)
	3 units	5 units

8.1 Repeatability

The results of duplicate determinations, carried out at different times, in the same laboratory by the same operator with the same apparatus on the same analysis sample, shall not differ by more than the above value.

8.2 Reproducibility

The means of the results of duplicate determinations, carried out in each of two laboratories on representative portions taken from the same sample after the last stage of sample preparation, shall not differ by more than the above value.

9 TEST REPORT

The test report shall include the following particulars :

a) the reference of the method used;

b) the results and the method of expression used;

c) any unusual features noted during the determination;

d) any operation not included in this International Standard, or regarded as optional.

Dimensions in millimetres

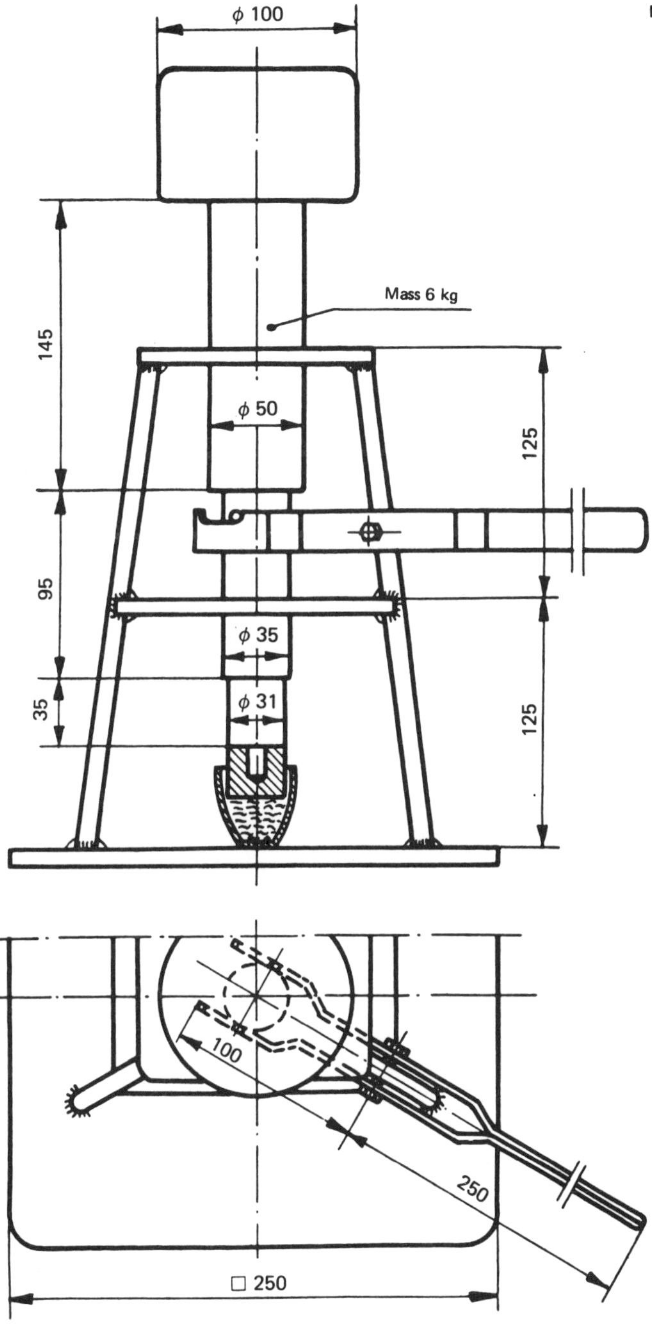

FIGURE 1 – **Press used for compressing the mixture of anthracite and test coal**

Dimensions in millimetres

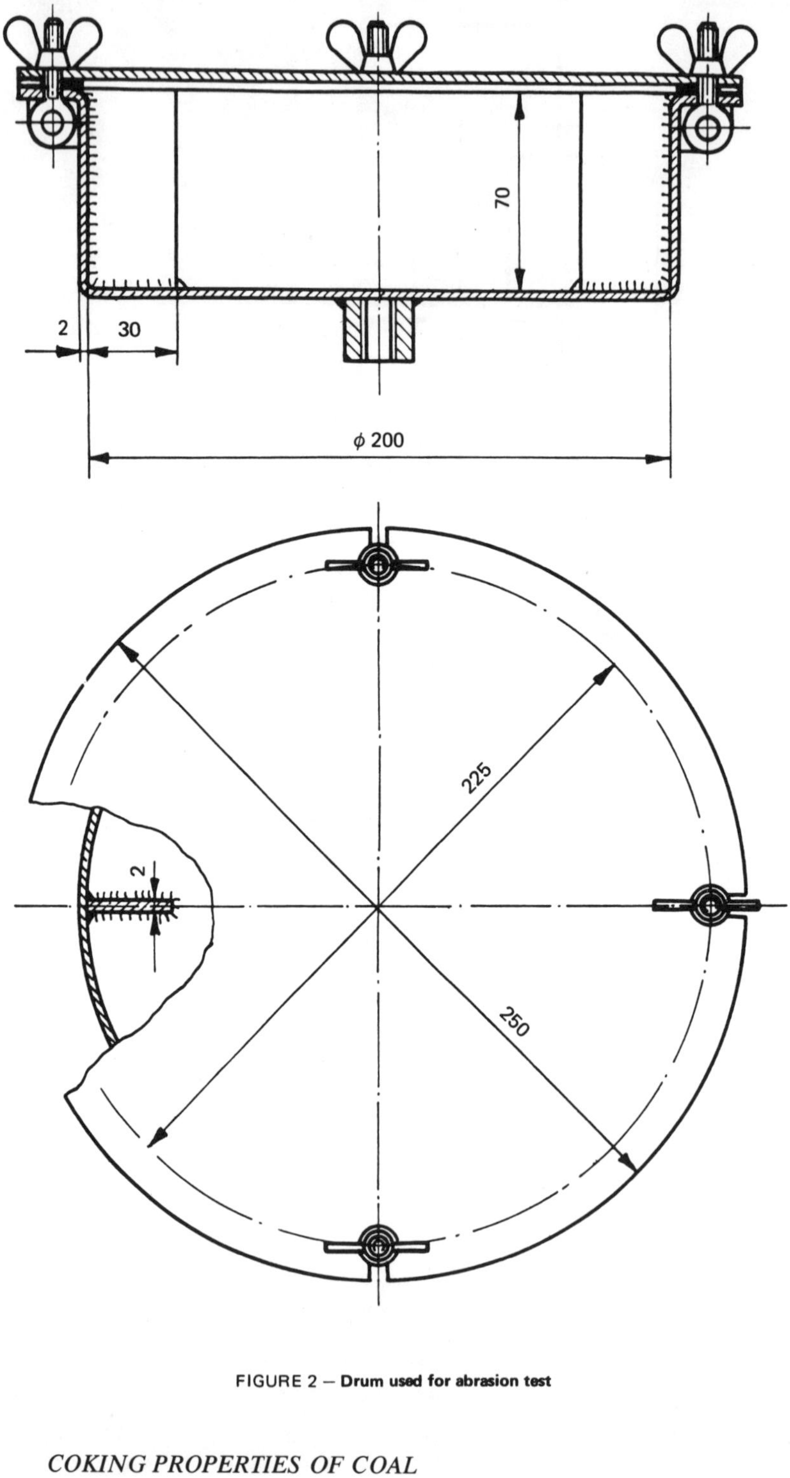

FIGURE 2 — **Drum used for abrasion test**

Appendix E

Second Draft Proposal for the Micum Test
ISO Standard TC 27/Secretariat-251, ISO(SFE)5005
ISO Technical Committee 27, Solid Mineral Fuels

Supplement

Chairman's Report on the Work of ISO Technical Committee 27 SC 3—Coke
Submitted by M. D. Sanna, France, Chairman of ISO/TC 27/SC 3
Dated July 20, 1976

This material is reproduced with permission from the American National Standards Institute, 1430 Broadway, New York, New York 10018.

International Organization for Standardization

ISO TECHNICAL COMMITTEE 27 - SOLID MINERAL FUELS

SECOND DRAFT PROPOSAL FOR THE MICUM TEST

Object

The object of the test is to determine the mechanical strength of coke.

Principle

Coke sieved over a specified size is subjected to treatment under standardized conditions in a rotating drum. The results of the sieve analysis of the treated coke are used as an index of the mechanical strength.

Apparatus

(i) Test drum

The test drum shall consist of a cylindrical drum, 1,000 mm in internal length and 1,000 mm in internal diameter, of steel plate of minimum thickness 5mm (as used, i.e. even after wear) and preferably of welded construction.

Four angle irons, each 100 mm by 50 mm and 10 mm thick and of the full length of the drum, shall be fixed to the internal cylindrical surface parallel to the axis of the drum and spaced at 90° intervals so as to divide the periphery into four equal sections. They shall be fixed so that the arms which are 100 mm long point towards the center, the other arms being attached to the surface and orientated in a direction opposite to the direction of rotation of the drum.

The drum shall have an opening in the cylindrical wall, for charging and discharging. This shall be as large as possible to facilitate emptying and cleaning of the drum. It is recommended that the size of this opening should be 600 mm (parallel to the axis) by 500 mm. It should be possible to close the charging hole by a curved cover of suitable thickness which can be fixed in such a way that when it is in place the inside is level with the internal surface of the drum. A rubber washer may with advantage be used on the cover to minimise loss of dust.

The drum shall be mounted in bearings by means of a stub axle attached to each end (i.e. the axle must not pass through the drum) to allow of its being rotated on its horizontal axis at a constant speed of 25 rev/min. The bearings shall be mounted on supports of height suitable to allow of convenient emptying and cleaning of the drum after a test. A suitably fixed sloping plate shall be fitted underneath and close to the drum to facilitate collection of the coke when it is discharged. It is desirable that the apparatus should be fitted with a revolution counter preferably with an automatic device to stop the drum after 100 revolutions.

(ii) Weighing machine

A weighing machine capable of weighing a nett quantity of coke of up to 50 kg. to an accuracy of 0.05 kg.

(iii) Sieves

Sieves, conforming in detail with the specification for sieving tests on coke, of aperture size 60 mm, 40 mm and 10 mm (round). A sieve of aperture 20 mm (round), will also be required if it is desired to test the product at this size.

(iv) Containers

Suitable light sheet-metal containers for the sample and the fractions sieved from the products. The largest container shall be suitable to contain 50 kg. of lump coke.

Sample of Coke

A minimum total weight of 150 kg. of coke of size over 60 mm. round aperture is required. This sample shall be collected from the gross sample of the consignment of coke.

Each determination for the Micum test shall be made on 50 kg. ± 0.5 kg. of coke, from the sample, weighed to an accuracy of ± 0.05 kg. The moisture content of the coke as tested shall not exceed 5 percent. The method of sieving the sample is described later.

Procedure

The 50 kg. sample of coke shall be carefully charged to the drum, thoroughly cleaned from previous residue, in such a way as to avoid breakage, and the cover put in place. The drum shall then be rotated at a constant speed of 25 rev/min over a period of 4 min ± 10 sec. The drum shall then be stopped, the cover removed, and the product carefully removed and sieved on the 40 and 10 mm sieves. If desired, the product may also be sieved on the 60, 20 mm. or other sieves. The cumulative weight shall be determined and recorded to minimise weighing errors in the cumulative percentage. Any difference between the sum of the weights of all the fractions and the weight of the coke charged shall be added to the weight of the -10 mm fraction. If this loss exceeds 0.7 percent of the weight of the coke charged the test shall be rejected.

Method of Sieving

The sieving of the sample of coke taken for the test and of the fractions in the products shall be by hand-placing, any piece of coke which can in some position be passed through a sieve aperture being considered as undersize.

Number of Tests

At least two tests shall be made on each coke. If the ranges of values obtained exceed

3.0 units in the percentage of coke over 40 mm or
1.0 unit in the percentage of coke under 10 mm

Then at least one further test on the coke shall be made and the mean values of the three or more tests taken.

Expression of Results

The mean results of the tests correct to one decimal place, shall be reported as follows:-

Mechanical Strength: Micum M40.

The percentage of the original coke remaining over 40 mm (round) after the test.

Abrasion Index: Micum M10.

The percentage of the original coke under 10 mm (round) after the test.

The percentages oversize on 60 mm, 20 mm, or other sieves may also be reported if desired.

The moisture content of the coke sample before test shall be stated in the report.

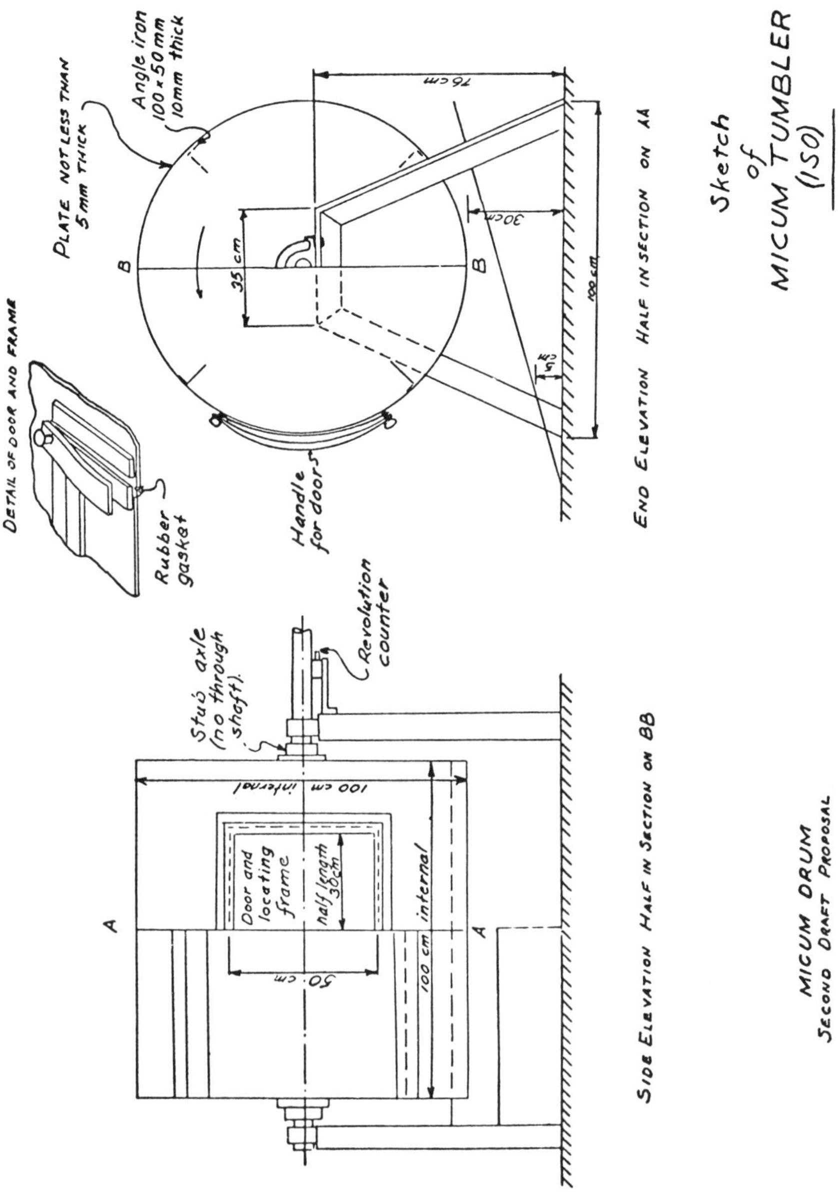

Detail of door and frame
Rubber gasket
Plate not less than 5mm thick
Angle iron 100 x 50 mm 10mm thick
B
B
35 cm
30 cm
76 cm
100 cm
5 cm
Handle for doors
End Elevation Half in section on AA
Sketch of MICUM TUMBLER (ISO)
A
A
Stub axle (no through shaft).
Revolution counter
100 cm internal
Door and locating frame
half length 30 cm
50 cm
100 cm internal
Side Elevation Half in Section on BB
MICUM DRUM
Second Draft Proposal

Appendix E Supplement

CHAIRMAN'S REPORT ON THE WORK OF
ISO/TC 27/SC 3 - COKE
(submitted by M. D Sanna, France, Chairman of ISO/TC 27/SC 3

Since the last meeting of TC 27 (Washington, September 1973) Sub-committee 3 "Coke" has met twice:

- in Paris on 4, 5 and 6 June 1974;
- in London, on 28, 29 and 30 June 1976.

These two meetings which were attended by about 20 delegates from 7 or 8 delegations were particularly encouraging and rewarding on account of the conducive atmosphere surrounding the discussions, the number and quality of the documents submitted, the results achieved with regard to international standardization and the general pattern emerging for future work.

1. Mechanical resistance of coke

In conformity with resolution 222 adopted at the last meeting of TC 27, Sub-committee 3 has reexamined the drum test standards R 556 and R 1881; at the meeting in Paris the Sub-committee instructed the Secretariat to prepare a draft combining both the MICUM and IRSID tests in a single text which was discussed and finalized at the recent meeting in London. The nature and advantages of this document, which has been forwarded to the Secretariat of TC 27 to be submitted to a postal ballot, are as follows:

a) the test uses the ISO drum described in R 556 and R 1881 (formerly the MICUM drum); it also uses the half-drum derived from the above for which both the degradation characteristics and the value and dispersion of the results obtained are identical.

b) the coke subjected to testing (> 20 mm instead of > 63 mm coke for R 556) corresponds to the material used in blast furnaces.

c) the dimension of the coke tested is determined by complete size analysis at three stages during degradation:

before testing, when knowledge of the size analysis of the overall sample enables granulometrically identical test samples to be prepared thereby reducing the dispersion of the results;

after 100 revolutions of the drum:

finally, after 400 further revolutions of the drum, a total of 500 revolutions (e.g. IRSID test).

In its present design the new combined ISO test constitutes a real step forward vis-à-vis its predecessors:

- both with regard to the procedure which has been made more stringent by the introduction of size analysis inspection;

- and with regard to the interpretation of the result where possibilities have increased.

2. Coke sampling

The standard 2309 benefits from two amendments approved at the recent meeting of SC 3 in London and regarding:

- sampling from stock

- sampling of small coke less than 20 mm in diameter.

These amendments have been sent to the Secretariat of TC 27 to be forwarded to the ISO Central Secretariat for circulation as a draft amendment (ISO/DAD) and for postal ballot.

Furthermore, with regard to the sampling of coke on large flow conveyors a significant experimental contribution has been made by the Japanese delegation who have been requested to prepare a draft supplement to the standard 2309 for discussion at the next meeting of SC 3.

3. Size analysis of small coke < 20 mm

The Secretariat of TC 27 has been requested to submit the amendments to standard 2325 agreed at the last meeting to ISO Central Secretariat for circulation as a draft amendment (ISO/DAD) for letter ballot.

4. Total moisture of the coke

Experimental contributions submitted confirm the weakness of the indirect approach and the validity of standards 2309 and 570.

5. Determination of the nitrogen - Kjeldahl method (ISO 333)

Experimental contributions submitted, in particular by the Japanese and UK delegations, show that this method provides results which are unsatisfactory, e.g. with respect to the combustion method, for certain types of coke considered to be slow to react and difficult to attack. It has been suggested to TC 27 to remove coke from the field of application of this standard. Studies are now entering the field of scientific research and it would be advisable to wait.

6. JIS drum resistance test

Sub-committee 3 is collecting data to enable the advantages of the JIS test to be assessed. Discussion is on the programme for the next session.

7. Proposals regarding future work

Some delegations have shown a certain amount of interest in the high temperature tests. These are still a matter for industrial research, but it would not be unrealistic to consider that a draft standard might be discussed within a relatively short space of time. Reactivity and resistivity do not seem to be so close to standardization.

Conclusion

The next session will see the completion of the usual type of standards having a technico-commercial nature or application; it may see the preparation of new standards with a practical objective, but with considerably increased help from science.

Appendix F

Japanese Industrial Standard Drum and Tumbler Test
JIS K 2151-1960

JIS Drum and Tumbler Test

6.2 Drum Test

6.2.1 Purport The drum in which the sample is charged shall be rotated for the specified cycle of revolution at the specified rate of revolution. Then the content shall be subjected to the sieving process through the specified number of sieves and the product retaining on each sieve and that which passed through the smallest sieve shall be weighed separately, reporting the result by the cumulative percentage to the original weight of the sample.

6.2.2 Sampling The sample shall be taken in accordance with the procedure described in 3 in JIS M 8811, with the exception that the number of increments is increased so as to contain not less than 40 kg of lump coke of about 50 mm size when the gross sample contains less than 40 kg of coke of the size. The gross sample shall be air-dried and screened (7) through the 50 mm sieve specified in 2. Size Analysis. Collect the sample which retains on the 50 mm sieve, reduce it by the increment reducing process or the like and take therefrom 10 kg of the sample.

Note (7) The shaking of the sieve shall be done in such a manner that most of the undersize is allowed to pass through the sieve in any position and the residue be hand-placed until no more will pass through it.

6.2.3 Apparatus (See Attached Fig. 5)

(1) Drum Test Machine The drum test machine shall consist of a steel drum of 1500 mm in inside diameter, 1500 mm in length and 9 mm in thickness. Six blades of flat steel, 250 mm in width and 9 mm in thickness, shall be fixed to the internal cylindrical surface, parallel to the axis of the drum and spaced at equal intervals. Six openings with covers shall be provided in the side wall for introducing and removing the sample.
During the test, the cover shall be rigidly fastened to the drum which is mounted on a horizontal shaft without passing through the drum and the drum shall be provided with a revolving apparatus so as to rotate at a speed of 15 ± 0.5 revolutions per minute by a proper electrically driven device.

(2) Sieves Three kinds of sieves of 50 mm, 25 mm and 15 mm openings specified in 2. Size Analysis shall be used. Two kinds of sieves of 38 mm and 6 mm openings shall, if necessary, be used in addition to the three kinds of sieves. All kinds of sieves to be used shall have square frames of 350 x 350 mm in area and 100 mm in depth as illustrated in Attached Fig. 4.

6.2.4 Procedure

(1) Weigh appropriate amount of the sample and place it gently in the drum.

(2) Tightly close the covers of the drum and subject the drum to a rotation of 30 revolutions for two minutes after which the drum shall instantly be ceased to rotate within the tolerance of 1/6 revolutions.

(3) Take the product out of the drum and subject it to the sieving process with the largest opening sieve at the top (8).

Note (8) The shaking of the sieve shall be done in accordance with the process as in 2. Size Analysis and especially with the sieves of 50 mm and 38 mm in size, it shall be done in such a manner that most of the undersize is allowed to pass through the sieve in any position and the residue be hand-placed until no more will pass through it.

(4) Separately weigh the coke retaining on each sieve and that which passes through the smallest opening sieve.

(5) If the sum of the weights for each size reveals a loss more than 1 %, the test shall be rejected and another test shall be made.

6.2.5 Number of Tests The drum strength test shall be made successively until three test results would agree within the tolerance given in 6.2.7.

6.2.6 Calculation The ratio of the weight of coke retained on each sieve to that of the total sample shall be calculated to one place of decimal in percentage respectively.

6.2.7 Tolerance The difference between the maximum and the minimum values obtained by three tests for the cumulative percentages retaining on 15 mm sieve shall be given as follows:

15 mm Min (%)	Tolerance (%)
90.0 Min	3.0
80.0 Min	5.0
70.0 Min	6.0
60.0 Min	7.0
Not more than 60.0	12.0

6.2.8 Designation The test results shall be reported in cumulative percentages by the mean of three test results calculated to one place of decimal as follows:

Total retained on	Cumulative percentage (%)
50 mm sieve	
25 mm sieve	
15 mm sieve	

Remarks: The above form of designation shall also be applied to the tests obtained by using the 38 mm and 6 mm sieves.

6.3 Tumbler Test

6.3.1 Principle The sample is introduced in the specified apparatus, which is rotated for the specified cycles at the specified velocity and the product is sized through the specified sieves. The product retained on each sieve and that which passes through the smallest sieve is weighed separately and the result is reported by the cumulative percentage to the weight of the sample.

6.3.2 Sampling The sampling shall be made in accordance with 3 in JIS M 8811.
In case the gross sample contains less than 30 kg of lump coke of the size 75 mm ([9]) to 50 mm, the number of increments shall be increased so as to contain more than 30 kg. The taken gross sample shall then be air-dried and sized ([10]) through 75 mm and 50 mm sieves specified in 2. Size Analysis. Collect the sample of the size 75 mm to 50 mm and subject it to the increment reduction process to take 10 kg of the sample therefrom.

Notes ([9]) The lump coke coarser than 75 mm shall be broken to 75 mm or smaller by prying apart at fracture cracks with a suitable tool, taking care not to drop or hammer the lump violently.

([10]) The shaking of the sieve shall be done in such a manner that most of the undersize is allowed to pass through the sieve in any position and the residue be hand-placed until no more will pass through it.

6.3.3 Apparatus (See Attached Fig. 6)

(1) Tumbler Machine The tumbler machine shall consist of a circular steel drum 914 mm in inside diameter and 457 mm in inside length made of steel plate at least 6 mm in thickness. Two equally spaced 50 x 50 x 6 mm angles shall be riveted longitudinally inside the drum. These angles shall be riveted to the shell so that the attached legs point away from the direction of rotation, thus giving a clear unobstructed shelf for lifting the coke. An opening shall be provided, preferably in the shell, for introducing and removing the sample. During the test, the cover shall be rigidly fastened to the shell and shall be so constructed as to fit into the shell in order to have a smooth inner surface. The drum shall be mounted horizontally on the shaft (38 mm in outside diameter) passing through the drum and be provided with an electric driven device which rotates the drum at the rate of 24 $\pm$ 1 revolutions per minute.

(2) Sieves For sieving the coke after the tumbler test, five kinds of square hole sieves with the openings as listed below shall be used.

50 mm, 38 mm, 25 mm, 12.5 mm and 6 mm

The dimensional requirement of the frame of each sieve shall be 350 x 350 x 100 mm as shown in attached Fig. 4

6.3.4 Procedure

(1) Weigh accurately the sample and place gently in the drum of the tumbler machine.

(2) Fasten the cover rigidly and rotate the drum for 400 revolutions (about 17 minutes) or 1400 revolutions (about 58 minutes).

(3) Remove the sample from the drum and sieve using five square mesh sieves ([11]).

Note ([11]) The sieving of the coke shall be made in accordance with 2. Size Analysis and, especially for 50 mm sieve, the sieve shall be shaken so that most of the sample is allowed to pass through it in any position and hand-place the residue until no more will pass through it.

(4) Weigh separately the coke retained on each sieve and that which passes through the smallest sieve used.

(5) When the sum of the weights retained on each sieve and that which passes the smallest sieve shows a loss of 0.5 % and larger, the test result shall be rejected and another test shall be made.

6.3.5 Number of Tests The tumbler tests shall be made successively until two test results would agree within the tolerance given in 6.3.7.

6.3.6 Calculation The ratio of the weight of coke retained on each sieve to that of the total sample shall be calculated to one place of decimal in percentage.

6.3.7 Tolerance The difference between two test results shall be 7 % on the basis of cumulative percentage for the size more than 6 mm inclusive.

6.3.8 Designation The mean of two tests results shall be calculated as follows:

Size (mm)	Cumulative percentage (%)
50 Min	
38 Min	
25 Min	
12.5 Min	
6 Min	

On designating the results, the number of revolutions of the drum shall be recorded.

7. Reactivity Test

7.1 Principle The sample is placed into a reaction tube kept at specified temperature up to the specified height and CO_2 gas introduced therein at the specified rate of flow. The rate of flow of carbon monoxide gas evolved is measured to obtain the volume of evolved gas, which is expressed as the reactivity of coke.

7.2 Sampling The sampling for the test shall be made in accordance with 3 in JIS M 8811. The air-dried sample shall be pulverized to appropriate size ([12]), reduced and sieved through two sieves 830μ (20 meshes) and 1980μ (10 meshes). About 100 g of the sample which retained on the 830μ sieve and passed through the 1980μ sieve shall be taken as the sample.

Attached Fig. 5 Drum Test Machine (Example)

Unit: mm

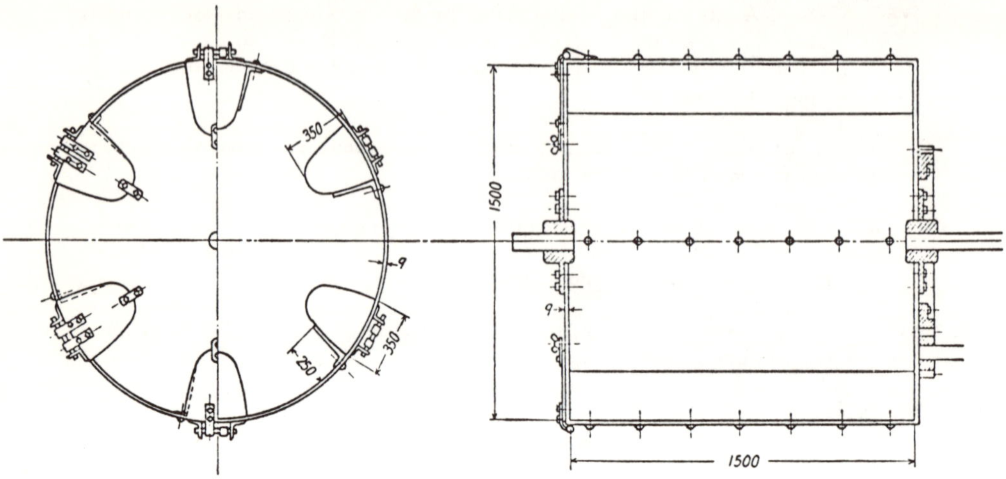

Remarks: The revolving apparatus is not shown.

Attached Fig. 6 Tumbler Test Machine (Example)

Unit: mm

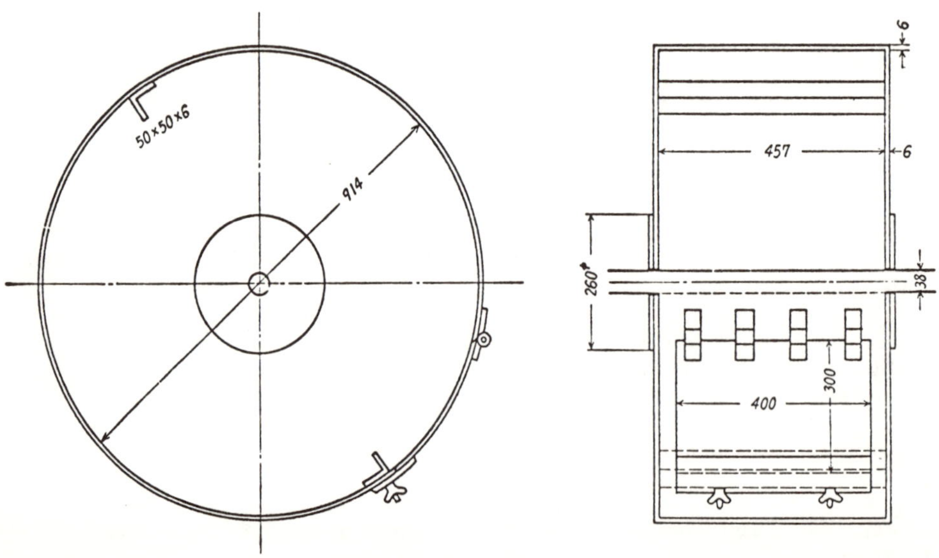

Remarks: The revolving apparatus is not shown.

Appendix G

Sole-Heated Oven Dilatometer for Measuring the Expansion
and Contraction Behavior (Volume Change) of Coals
ASTM D-2014-71

Reprinted with permission from the *Annual Book of Standards*, copyright American Society for Testing and Materials, 1916 Race Street, Philadelphia, Pennsylvania 19103.

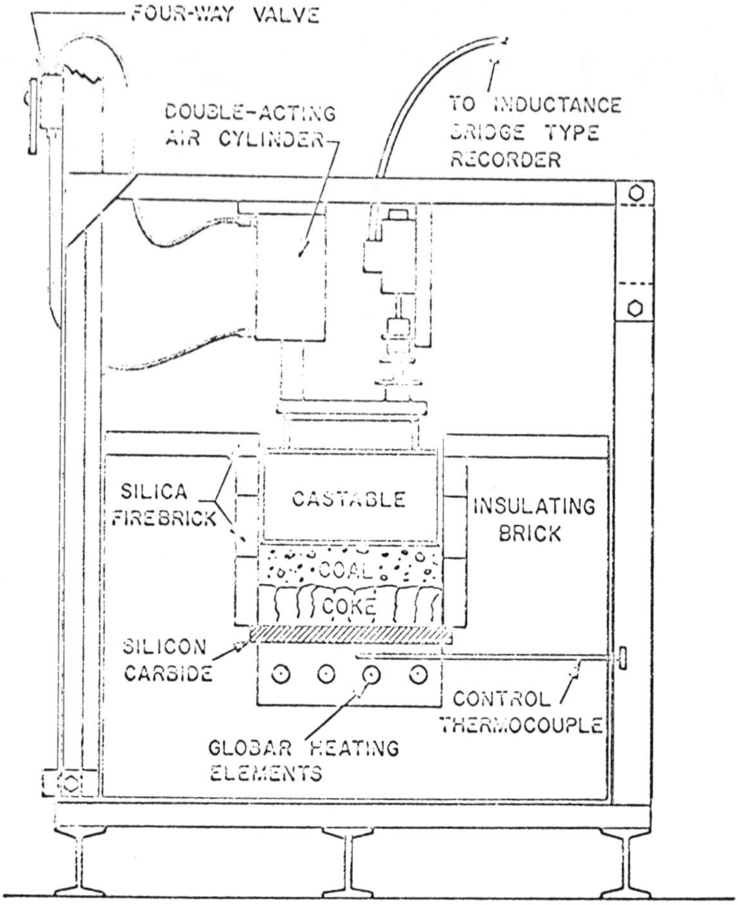

Sole-Heated Oven Dilatometer For Measuring the Expansion and Contraction Behavior (Volume Change) of Coals

Appendix H

Adjustable-Width Oven Illustrations

From R. W. Schoenberger and B. R. Kuchta, AIME, Iron Making Conference, Toronto, Canada, *Design and Application of an Adjustable Width Movable-Wall Type Oven*, 1969.

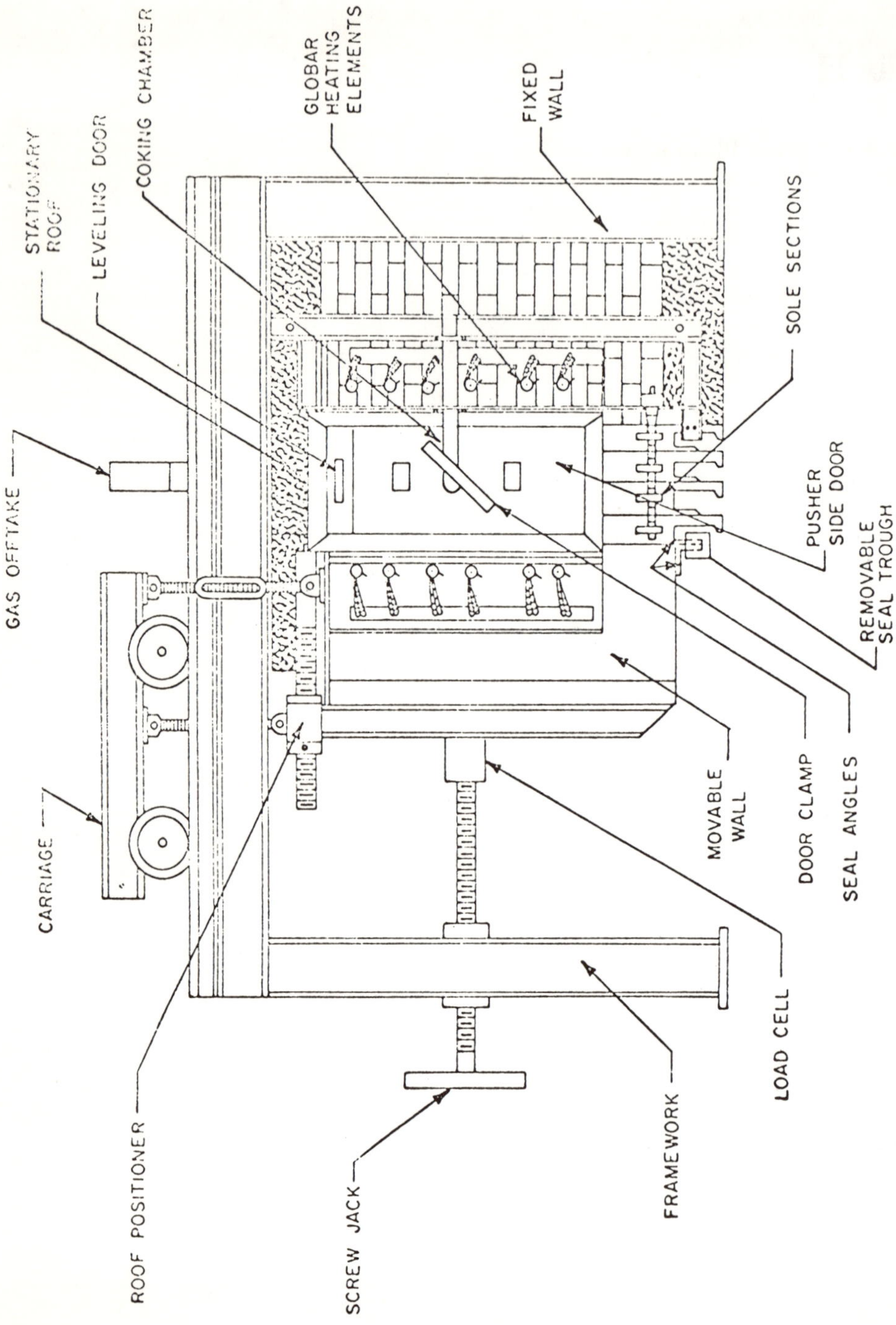

ADJUSTABLE-WIDTH OVEN

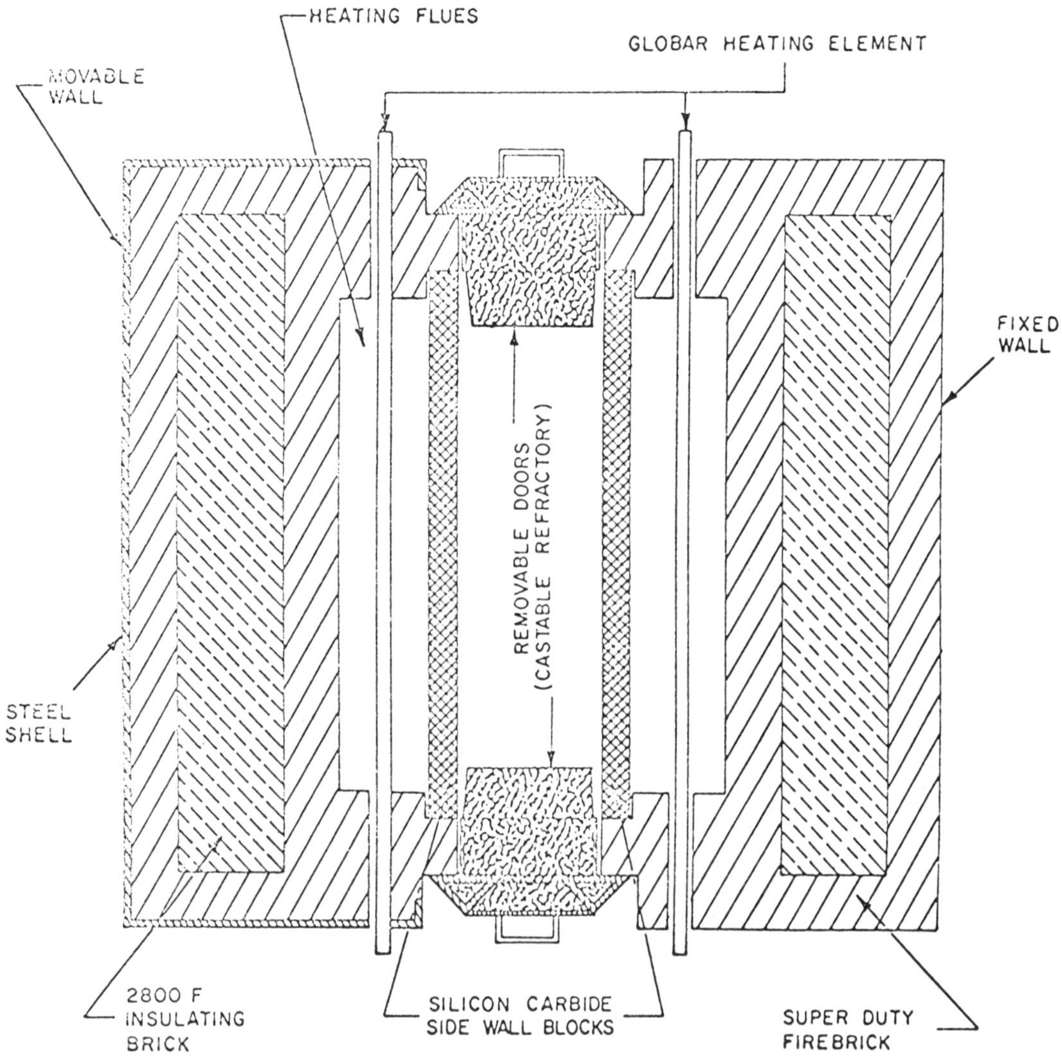

ADJUSTABLE-WIDTH OVEN, TYPICAL BRICKWORK ARRANGEMENT

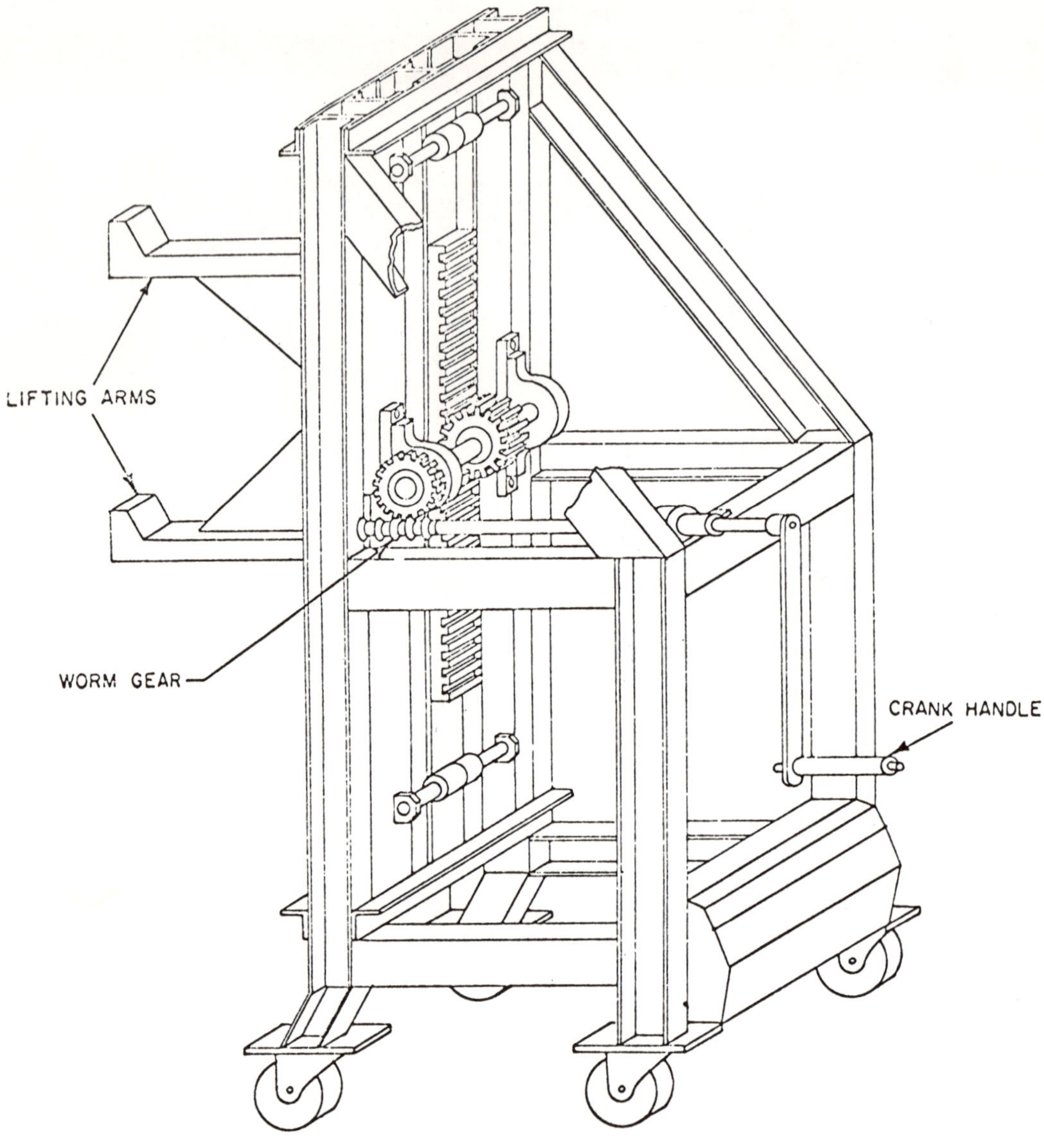

OVEN-DOOR EXTRACTOR

Index

About the author

Raymond E. Zimmerman has devoted his life to the mineral field, especially in coal utilization and beneficiation. His experience is worldwide, and he is called upon for consultation in many countries.

He has his B.S. and E.M. degrees in Mining Engineering from the College of Mineral Science, Pennsylvania State University. He is a registered Professional Engineer in both the U.S.A. and Canada. He is a member of the honorary societies of Sigma Gamma Epsilon (Mining), Phi Lambda Upsilon (Chemistry and Physics), and Sigma Xi (Graduate Research). He is mentioned in *Who's Who in Engineering, Who's Who in the Midwest,* and *American Men in Science and Education.*

Among the various technical organizations of which he is a member are the American Institute of Mining and Metallurgical Engineers (AIME), the Canadian Institute of Mining and Metallurgy (CIM), and the Mineral Processing Society of Great Britain.

He was at one time professor and head of the Department of Mineral Preparation at Pennsylvania State University. He has held many responsible positions in his field, including director of research, Pittsburgh Coal Carbonization Company, and chief preparation engineer, United States Steel Corporation. His present position is vice president, Paul Weir Company, where he specializes in coal carbonization and coal preparation. He is the preparation editor for the magazine *World Coal.*

He is author of many technical papers, and a co-author of the AIME book *Coal Preparation.*